THE IDEA OF WASTE

THE IDEA OF WASTE

ON THE LIMITS OF HUMAN LIFE

JOHN SCANLAN

REAKTION BOOKS

Published by
Reaktion Books Ltd
Unit 32, Waterside
44–48 Wharf Road
London N1 7UX, UK
www.reaktionbooks.co.uk

First published 2025
Copyright © John Scanlan 2025

All rights reserved

No part of this publication may be reproduced, stored in a retrieval system, or transmitted, in any form or by any means, electronic, mechanical, photocopying, recording or otherwise, without the prior permission of the publishers. No part of this publication may be used or reproduced in any manner for the purpose of training artificial intelligence technologies or systems.
Printed and bound in Great Britain
by CPI Group (UK) Ltd, Croydon CR0 4YY

A catalogue record for this book is available from the British Library

ISBN 978 1 83639 034 3

CONTENTS

Introduction: Waste Is Life Plus Minus

For nine days in February 1968 the streets of New York City looked like a vision of the future. Or so thought Abbie Hoffman, figurehead of the Yippies (Youth International Party). A strike by 10,000 sanitation workers meant that household waste – trash, garbage, refuse and whatever other names it had – was taking over the streets. The *New York Times* reported that just three days after collections ceased, 'many once-clean sections' of the city now looked 'like a vast slum as mounds of refuse grew higher and strong winds whirled the filth through the streets'.[1] It was not hard to dredge up the sense of looming apocalypse. Hoffman was then writing a book (published as *Revolution for the Hell of It* in 1968) whose style and language would reveal a preoccupation with waste. It was, for Hoffman, a handy metaphor for life in that time and place ('The book is just another part of life,' he writes; 'A kid tells a story, a trigger flashes in my head, and in goes another page of garbage'), if not emblematic of his own struggle against the system. 'As I write this,' Hoffman noted:

> 100,000 tons of garbage are piled up on the streets of New York. I have a vision of the country being totally inundated under this massive garbage pile. Future historians would write that America was destroyed by a nuclear attack when in actuality the people just stopped picking up their trash.[2]

Chief among the representatives of authority who routinely clashed with the Yippies was the New York City Police Department, whose officers seemed to take pleasure in harassing young people who liked to be out on the streets after dark. But the youth found their own way to fight back, using what was lying around on the streets. 'In ten minutes Second Avenue was covered in a sea of garbage,' wrote Hoffman following one altercation with the police.[3] The message to the police was simple and literal: 'you mess with us, we mess with you.'[4] Indeed, infractions against Hoffman's sense of natural justice often seemed to involve as a response the dumping of trash in places where it would be most conspicuous. In 1964, when Black civil rights activists complained about the irregular nature of refuse collections in their neighbourhoods, he hatched a plan to take all their garbage and simply dump it right in front of City Hall.[5]

Environmental activism around questions of pollution would eventually project the problem of urban waste as something symbolic of a looming ecological breakdown, but at this point in time, aside from anarchist environmentalist Murray Bookchin and his followers, there was very little political activism that was motivated by the problem of waste.[6] For Hoffman, on the other hand, the reality of living with urban refuse represented a spectre of disorder that was welling up just below the surface of everyday life. This intuition allowed him to use the attention that was being paid to waste as a result of the refuse strike as a prop for political purposes, specifically to outrage the refined sensibilities of those who didn't have to deal with the poverty and social problems of Manhattan's Lower East Side, where he lived. February 1968, during the strike, was an opportune moment to make a topical protest. Despite pleas from the city authorities that residents should carefully bundle, crush and secure their trash so it would not spread all over the streets, refuse continued to spill onto the sidewalks and roads. In some places, the *New York Times* reported, 'mounds were higher than automobiles parked at the curbs.'[7] The protest targeted a cultural exchange event taking place at Lincoln Center, then a new home for the performing arts in Manhattan. 'March to Lincoln Center. Bring Your Own Garbage. Let's Trade It for Their Garbage – Even Steven', read the leaflets. Hoffman recalled: 'About thirty of us walked from our neighborhood through the streets of Manhattan to newly-opened Lincoln Center with our bags of garbage and dumped them in the courtyard fountain, scattering in every direction when the cops chased us.'[8]

The moment was captured in a 1968 home movie when Hoffman joined with fellow anarchists and sometime foes, who went by the charming name of 'Up Against the Wall Motherfuckers', to plot their assault on the patrons of Lincoln Center as they exited that night's opera performance. The shaky black-and-white footage follows the protest as it proceeds via subway to its destination. In between the chaotic sounds of slogans loudly declaimed, the banging of makeshift drums and a random flute sounded by an anarchist Pied Piper, a voice cuts through the noise, reminding all that this waste problem was about more than it seemed. 'It'll be a cultural exchange,' says the voice, 'garbage for garbage', and 'America turns the world into garbage, it turns its ghettos into garbage . . . we of the Lower East Side have decided to bring this culture to Lincoln Center to where it belongs . . . we're just gonna smear reality all over them.'[9]

What makes the stuff of waste so potent, as Hoffman realized, is that it is at once very familiar to us, yet at the same time its afterlife is mostly obscure or mysterious. It is either hidden from us, or we ensure we remain well away from it. Perhaps it is only on those occasions when we find waste impossible to ignore that we are really aware of it. Such has been the case during similar garbage strikes in other places, which have revealed the shockingly prodigious multiplication of the stuff if it is not diligently managed.

That is the power of waste when it is visible. But waste is also much more than the material leftovers of our lives.

The critic Raymond Williams once wrote that 'culture' was one of the most complicated terms in the English language. A look at the *Oxford English Dictionary* reveals that 'waste' might be another word that matches it in this respect, running to ten pages (or 26 long and densely packed text columns) containing definitions of this seemingly familiar word that reach back as far as written English can be found, and tracing its arrival in English from older Latin and French: *weste*, *vast*, *wast* and so on.[10] And that's not counting its many synonyms or alternates: garbage, junk, trash, filth, ordure, excrement, refuse, rubbish, litter, wilderness, desolation, desert, emptiness, and the numerous compound words and terms it has spawned (of which the familiar wastebasket and waste disposal are just two).

The reality of waste today, in a sense, encompasses whatever all these words refer to and whatever images they conjure in the imagination, which is to say that our language describes the many variations, configurations, modalities and familiar forms that make waste something familiar to us. The presence of waste may additionally be intimated as much as it is visible or otherwise apparent in smells or sounds (is noise pollution a form of waste?). Regardless of our tendency to think of waste as something material, it is in truth more of a human relation than it is any physical or material entity, or particular kind of thing. On eBay today, for instance, there is a thriving trade in all sorts of what many would regard as junk. Not merely second-hand goods, but broken, imperfect and malfunctioning goods. If there is clearly a trade in such stuff, is it junk or not? And if that kind of waste is very much relational in nature and shares in a subjectivity with many other wastes, there is also, of course, the waste that seems beyond redemption: perhaps where it represents a level of toxicity, pollution or danger that makes it universally recognized for some hazardous property. What we might say is that what we call 'waste' is something that is always becoming itself; it fluctuates as a human relation to the world (or things of the world) between polarities of absolute positivity and negativity and, at times, as in the case of eBay junk goods, represents an intermediate condition whose place in the positive–negative spectrum is subjectively determined. There are popular maxims that underline this fact: where there's muck, there's brass; one person's trash is another's treasure, and so on.

Conceptually, waste is a phenomenon of indeterminacy or in-betweenness. What it represents within value hierarchies at one moment, or in one circumstance, may be reversed. Excremental waste serves as nourishment for plant life and has also been utilized for its cleansing and purifying properties, and even as fuel in some public transportation systems. It was once even sealed up in tin cans by an artist who matched the value of this work to the

Piero Manzoni, *Merda d'artista* (*Artist's Shit*), 1961, demonstrates how the determination of 'waste' is contextual and subjective.

gold standard, turning what might be thought of as the most base form of matter into a source of value, but only within an entirely conceptual realm of understanding that is known as the art world. There is, then, a certain consciousness of waste, which makes itself known through a language that in turn reveals how, in order to exist as a phenomenon, waste takes form as something material or ideational that, first of all, has to be thought into existence. Cultural norms and structures – knowledge, technologies and psychosocial relations – also clearly 'create' and obscure waste. So, how might we conceptualize something that is at once so familiar yet also so obscured from the reality we produce?

In 2003 William J. Mitchell wrote a book titled *Me++: The Cyborg Self and the Networked City*, which revealed some of the ways that humans had been extended into a vast array of communication networks through the new digital technologies of the early twenty-first century.[11] Thinking about the idea of waste, likewise, involves seeing life through the means and methods by which we project ourselves into and onto the world and give meaning to our lives. These means and methods are the subjects of the six chapters of this book. If all those ways of adapting to the world – of attempting to dominate, exploit or control it – positively expand the reach and power of the human, then waste, which is an outcome of all human endeavour, negates these efforts. So, we might say that what waste represents is *life plus minus*. Waste raises the spectre of finitude, or the limits of human life. As an idea, waste has always existed between life (or what is life-giving, life-sustaining, life-enhancing) and death (or what represents a threat, a danger to health or a disaster of one kind or another). It points us towards a consideration of how everything we do holds those two opposites in balance, or not. Life plus minus. In that sense the idea of waste reveals a fluid or shifting and indeterminate category of phenomena, and whatever it is – or however we recognize and describe it – it is something that inheres in or attaches itself to everything we do at all times, whether or not we are conscious of it. This is one reason why it has been thought of, described and imagined in so many ways.

To the extent that we can sketch a history of waste through its manifestations, or through the accounts of various protagonists who have described its effects or tried to transform it, or through events that can help to inform our understanding of it as a shape-shifting phenomenon, what is revealed are transformations in what is ultimately *an idea*, an object of human consciousness. In that sense, this book is about what waste is and what it has been; the forms it has assumed and the challenges it has presented. It is about how we have lived with waste, made use of it as a thing or idea, and dreamt of escaping or conquering its negative effects once and for all. Finally, it is about how waste never disappears but rather only proliferates in new forms, contexts or as a future threat that human life, culture and society provide the grounds for.

A view into the main tunnel of London's 25 kilometre (15½ mi.) Thames Tideway Tunnel or 'super sewer', completed in 2024, proof of the never-ceasing battle against waste.

1
Matter: Sewers, Filth and Sanitarians

Before humans lived in settled communities and in the places that they came to belong to, they left behind any waste they created that was not reused as they moved on in their itinerant way. Those today who choose to turn their back on the modern methods and conveniences that allow us to push waste to the margins of consciousness may opt to embrace the waste they create, and express their desire to live more sustainably by choosing a compost toilet.

In the words of *The Humanure Handbook*, systems such as compost toilets are 'attractive to anyone seeking a low-impact lifestyle, and who is willing to make the minimal effort to compost their organic residues'.[1] No flushing, no forgetting, only the presence of waste as the necessary and indestructible symbol of life, and the knowledge that whatever else you are, you are your waste. At the other extreme, the place of waste in our highly rationalized, technocratic societies is a precarious one, dependent on never-ceasing management and control, the functioning of infrastructure systems invisible to us, and the labour of those who would step into the hideous spaces that separate cosy domesticity from the disgusting reality of the stuff that flows under our homes, villages, towns and cities.

Flushing Is Forgetting

'It looks like a truffle, it smells like a rotting whale, but there's something oddly alluring about the lump of Whitechapel fatberg that I've found myself cradling,' wrote a *London Evening Standard* investigator when confronted with one baby-sized, caked portion of the fats, oils and other ill-considered matter that had been flushed down the sinks, toilets and drains of London. At some point following the disappearance of this stuff into the vast network of channels through which it (mostly) flows, it will meet other forms of matter, where, depending on the circumstances, it may end up congealed into a large mass to form a blockage. In this case, the result was a single large and concrete-like 'behemoth' (like a living thing, a monster of staggering proportions), reported to be 'the length of two football pitches and the weight of 11 Routemaster buses'.[2]

The urban underworld has always had its initiates, but none more knowledgeable than the sewer workers who embrace this dark, malodorous realm; a meeting place for human and non-human life that most of us would never dream of entering. Wading through the cornucopia of unwelcome objects and fluids that find their way into the sewer – which might include needles, bones, shards of glass and dangerous chemical fluids – to deal with the Whitechapel fatberg, the operatives of Thames Water had gas masks at the ready just in case they were caught out by an unexpected gust of noxious fumes. Waste matter stinks, but none so much as that which finds its way into the enclosed channels of a sewer system. Sewer workers also had one eye on the lookout for other threats: oversized rats or unexpected tidal waves carrying excrement, which might at any time be flushed through by torrents of rainwater entering the system further back in the network of drains and tunnels. They were battling what had been characterized in the media as monstrous matter: 'an evil thing lurking beneath the streets'.[3]

In recent years these monstrosities have been named and identified. They are fatbergs, as alluded to above. Public information posters warning of the threat they pose to commodious living above ground appear on bus shelters and on the walls of underground rail networks. The Museum of London staged an exhibition of some of this matter, recovered from the city sewers, as part of an effort to shock the public into thinking before they flush. But the fact is that these fatbergs now seem to be finding a home beneath the streets of our cities, blocking the flow of water through the principal system of drainage via which the most basic of the wastes that we make can be banished. When the aforementioned London fatberg got stuck, homes in the vicinity of the blockage found themselves facing a backwash of excremental soup, pushing its way back up and into the domestic sphere. Like something out of a horror film, basements and ground-level rooms began to flood with raw sewage that kept arriving at the point of blockage – the fatberg – but had nowhere else to go except back where it came from; one of those signs, often unnoticed or ignored, that no waste ever really disappears. It is a story that reveals another truism about much waste: that it remains hidden only as long as the infrastructure and other systems that deal with it don't break down.

Other than the most determined of explorers chasing some special knowledge of what becomes of the effluvia of daily living – the ordure and other evacuations that we absent-mindedly flush away (along with whatever else enters through street drains and finds itself flowing invisibly beneath our feet) – few would seek out the opportunity to visit one of these dank and pungent subterranean channels. Places that become the destination for all manner of discards and into which some of the most basic truths of life – that all in the end is filth and decay – are pushed to the margins of consciousness. Some, however, embrace the opportunity to enter these underworlds. This was certainly true when the sewers of London and Paris were first constructed in the nineteenth century and seen as engineering wonders: 'physical manifestations of new patterns of water usage, bodily hygiene and the progressive application of new advances in science and technology'.[4]

Setting up his equipment in the tunnels of the celebrated Paris sewers, Nadar (Gaspard-Félix Tournachon), the pioneering nineteenth-century photographer, was able to reveal a world that was never seen by most people. Whatever his interest in the sewers was, Nadar was also interested in exploring the possibilities of the camera under closed conditions where electric lighting would have to be used. This fact pushed him towards recording more of the specific details of various underground places – sewers, catacombs – with the camera becoming a forensic instrument for seeing further and deeper into the unknown.[5] As Walter Benjamin would later write, Nadar's sewer photography raised him above his contemporaries as an exponent of what was the newest medium for seeing and recording reality, no matter how hideous, rather than just recording what was already there for all to see in the world above ground. 'For the first time,' Benjamin stated, in the hands of Nadar, 'the lens was deemed capable of making discoveries.'[6] The new spaces that Nadar revealed made the conventional uses of photography elsewhere at the same time seem unimportant. But as well as his expeditions underground in 1861 that were, in a sense, recorded for posterity in the images that we can look at today, Nadar recounted in writing the experience of journeying deep into the sewer. Aside from the workers who pushed him through the sewer in a cart, no one had entered the tunnels for sixty years, he claimed. A sense of mounting anxiety is clear in his descriptions of what followed. 'The high boots of our escorts clap in the horrible liquid, on the submerged sidewalks,' he wrote, the water rising up over their knees, then up to waist level, the threat of immersion in filth a real possibility: 'everything around us streams, puddles, flows, drips, oozes. The place has become totally sinister:

by the thick miasmas that float, our dimming lamps seem to be fainting, ready to go out.'[7]

Nadar made it out of the sewer with his camera and the pictures that would help to seal his reputation as an innovator in the new medium. His endeavour serves as one example, perhaps, of the fact that, out of our wastes and wastelands, if not out of the reconfiguration of waste matter itself, there often emerges a new thing, a new creation or possibility.

Another explorer who recorded a journey into this hidden realm was the London journalist John Hollingshead, who had developed a reputation for reporting on the most abject aspects of Victorian London in outlets such as the *Morning Post* and *Punch*, and whose journey into the sewers produced an 1862 book titled *Underground London*. 'Feeling a desire to inspect a main sewer almost from its source to its point of discharge into the Thames,' he wrote:

> I applied to the proper authorities, and was obligingly told that they had not the slightest objection to gratify what they evidently thought a very singular taste. I was even asked to name my sewer. They had blood-sewers (a delicate article) running underneath meat-markets, like Newport Market, where you could wade in the vital fluid of sheep and oxen; they had boiling sewers, which were largely used by sugar-bakeries . . . and they had sewers of different degrees of repulsiveness.[8]

The new nineteenth-century sewer systems were places of fear and disgust, to be sure, but also of fascination. Their existence tells us that at the most basic or elementary level, dealing with the wastes we flush away routinely and perhaps without any thought is a complex matter. As we will see in what follows, in its many guises waste actually finds a place in a variety of undergrounds

A visit to a newly opened sewer tunnel at Strasbourg, France, 1858.

that have been built by human societies: from sewage and railway systems to communications infrastructure and the deep geological burial of nuclear waste, and even those subterranean spaces that today may contain the 'cloud' servers that manage and transform what we might term 'data waste'. The significance of these spaces is in the effect that their existence has on our consciousness of waste. There is a logic of waste management that can be understood in terms of the well-known phrase, 'out of sight, out of mind'. When the regular functioning of this infrastructure is interrupted, what we are able to glimpse is another reality: an alternative domain of causality that we do not manage and do not control and which therefore seems to take on a life of its own. It is only on such occasions that we are truly aware of this hidden realm of waste, and that the reality of those drains, tunnels and the fluids that course relentlessly below can no longer be suppressed.

Out of Sight, Out of Mind

The earliest and most noteworthy success in channelling the tide of human excremental waste and other forms of refuse underground (and out of sight) was executed by the Romans when the legendary Cloaca Maxima, or great drain, was enclosed in the third century BC. Prior to this, an open channel through which a stream carried away various wastes had been initiated by Lucius Tarquinius Superbus, the last king of Rome, some three centuries before it was made into a tunnel. The Cloaca would ultimately become one of the most durable symbols of Roman civic virtue, if not of Rome itself, seeing as it outlasted many other structures that had defined the ancient city. Its status was such that the construction of the nineteenth-century sewers in the modern cities of London and Paris would see them compared as feats of engineering to Rome's great drain. In fact, engineers who were then building these great sewer systems that would become emblems of modernity followed archaeologists into recently cleared sections of the Cloaca Maxima in order to examine more closely its design and the materials it had been made from. What they found were 'immense blocks of peperino [volcanic rock] put together without cement', which formed a tunnel 5 metres in height and 4 metres wide (16½ × 13 ft) in a section that ran between its opening at the river Tiber and the Forum, which by the nineteenth century was the landscape of architectural ruins – the decaying fragments of the ancient city – that still exists today as a tourist attraction.[9] In effect, the Cloaca Maxima was an example of the way that innovations in waste management usually arise as adaptations to unexpected contingencies or new problems or circumstances. It transformed a stream that had run through the city into a channel that would more or less invisibly transport waste matter beneath the city and out to the river Tiber, where it would be spewed out and carried further out of sight by the movement of the tides.[10]

Today the Cloaca can be seen opening out below another ancient landmark, the Temple of Vesta, presenting a scene that would be captured in illustrations and guidebooks in the eras before and after photography. It is historically significant that what began with an open channel was first concealed by walls along its sides and then effectively pushed below ground level over time as the surface of the city rose (built up on centuries of accumulated matter). Once new layers of matter had reinforced the channel, fully enclosing it eventually became possible, thereby banishing from sight, if not smell, the wastes that flowed beneath. And as would be the case much later with the underground tunnels and drains of Paris and London, its concealment had the effect of making it an object of curiosity, providing it with the allure of a place that held truths or secrets about the life of the city.

In the case of the Cloaca, explorers of its interior would have perhaps marvelled at what Livy, writing in the first century BC, described as a monument without match in his own time.[11] Pliny celebrated the strength and durability of the drain in his *Natural History* (AD 77), noting that among the many impractical and costly building schemes that had seen Rome's rulers squander great quantities of wealth, the sewer was the 'most noteworthy achievement of all'.[12] The seven rivers that flowed downwards through Rome – 'like mountain torrents', Pliny wrote – were directed into the Cloaca Maxima in order that their waters could help to sweep away whatever waste material was put in their way. At times this created a spectacle that was intensified in its power when rainwater inevitably drained into the sewer:

> Sometimes the backwash of the Tiber floods the sewers and makes its way along them upstream. Then the raging flood waters meet head on within the sewers, and even so the unyielding strength of the fabric [of the tunnels] resists

> the strain. In the streets above, massive blocks of stone are dragged along, and yet the tunnels do not cave in.[13]

Indeed, for seven hundred years, Pliny notes, when the ground above was periodically shaken by earthquakes or buildings were brought down by fire, the sewer system 'remained well-nigh impregnable'.[14]

The French writer Louis Simond, a visitor to the city in the early nineteenth century, noted that where the mouth of the sewer let out into the Tiber, it was as if the tunnel deliberately projected itself out at that point as an invitation to modern explorers, 'so as to allow boats to penetrate into it as they did in ancient times'.[15] While the interior at that time was 'choked up', according to the account given by Simond, it was said that a clear stream of water that continued to trickle through was nonetheless drunk by city residents, serving as a purgative and diuretic during the summer months.[16] It sounds as much a story about something that had attained near mythical status as the greatest achievement of Rome as it does the truth. But if the Cloaca represented civic life as a means of seemingly conquering these inevitable wastes, which remains a task that never ceases, it is equally of interest to note that today the places where tourists might catch sight of the stamp of the ancient republic – SPQR (Senatus Populusque Romanus, or the Senate and People of Rome) – is on the surfaces of the apparently ancient waste infrastructure, those four letters being visible above ground on street furniture, 'from manhole covers to rubbish bins'.[17] The fact that these emblems of the city's ancient infrastructure, which are here presented as an aspect of a contemporary heritage culture and may be mistaken by some for the original, nonetheless perhaps illustrates another truism about waste that this book will explore: namely that no form of waste, once created, revealed or subject to control, ever really disappears.

The contemporary counterparts of the ancient drain covers are those found everywhere today on pavements and on roads, including those installed more recently in London by Thames Water, which now sometimes commemorate the defeat of some great fatberg. But waste, and the prodigious amounts of it that are produced in contemporary society, is still mostly hidden from us. The same, of course, is true of the infrastructure and waste management practices that keep it that way, whether we refer to its existence as a visible thing or as something embodied in the people whose labours remove it as a presence from our everyday lives.

As Pliny noted, we may justifiably look in awe at the fact that these constructions that we know to form lowly and disgusting places remain so structurally sound over such long periods of

Drain cover in present-day Rome, bearing the insignia of the ancient republic, SPQR – a modern link to the first known covered sewer, the Cloaca Maxima.

time. During his journey into the Paris sewers, Nadar was struck by the appearance of certain openings, especially those where several channels met and which were constructed like large domes or amphitheatres; clandestine spaces, he thought, that perhaps might 'offer quite useful points for the concentration of forces in some contingency'.[18] The fact is that should all else fail and the world above ground begin to crumble – whether through acts of destruction or simply through neglect – it will be the underworlds of our cities that remain relatively intact. This perhaps explains why, at the height of rising Cold War tensions in the early 1980s, when the threat of nuclear Armageddon seemed ever-present in the public mind, post-apocalyptic scenarios led British planners to develop strategically located underground facilities that would allow government to continue functioning regardless. They created spaces where ministers and civil servants would be sent to ensure that the nation – or what was left of it – could continue under a semblance of order and control, led by those who would be able to live and move underground.[19]

The first and most elementary of such subterranean worlds, though, belongs to the management of waste, which of course is also about the control or containment of looming disorder. In a newspaper report from 1925 concerning the construction of a new sewer in Manchester, England, which echoed Pliny's eulogy to the Roman sewer, it was noted that almost everything else that one might point to as examples of civic pride, in what was a celebrated era of great achievements in civil engineering and architecture, actually paled in comparison to these vast tunnels that were basically designed to carry away stuff that would otherwise bring life to a halt.[20] 'These tremendous subterranean undertakings may very easily be the most lasting memorials of our cities,' the unnamed reporter of the *Manchester Guardian* opined, noting also that 'Bridges and temples threaten to collapse around us at this very minute, but great engineering works carried out eighty feet and more below street level should have every chance of challenging Time with the success of their great Roman type and forerunner.'[21]

Thus, with the Cloaca Maxima, what the Romans also invented was the idea of the city expanding its forms of life and finding a way to maintain order through concealment: push it underground and out of sight. As urban settlements became, over time, more populous and societies developed in more technologically complex ways, the increasing number of support systems required to sustain life also had to be pushed out of the way. As the architectural critic Lewis Mumford implied in *Technics and Civilization* (1934), there was a distinct psychosocial aspect to this. Moving what we might now think of as the necessary wiring and plumbing of life out of view was something forced upon societies in order to avoid a sensory overload, even if in burying this other side of life all we were doing was hiding a barely controlled array of vulnerabilities so that we could continue life above ground without thinking too much about how easily it all might fall apart.[22] Out of sight, out of mind. And so, if the urban world was thought of as an organism, Mumford tells us, with the shifting and breathing life forms kept going by what lay concealed under the surface of the city, it came to pass – 'precisely because there are so many physical organs, and because so many parts of our environment compete constantly for our attention' – that they had to be moved into underworlds that now lie beneath the ground.[23] What Mumford really meant was that, in hiding away the subterranean city, 'the machine' – as he termed technologized societies – performs a civilizing function, much like social codes of interaction and manners do with regard to the maintenance of human relations in daily life between people.[24]

Written in the mid-twentieth century, Robert Daley's *The World beneath the City* plunged deeper and in more detail into what it took to keep things looking ordered above ground, preferring to think of our invisible undergrounds as expanding and 'unseen roots' of modern urban life, which at once nourish the city and represent the growth of 'incredible complexity and awesome power'.[25] Consider this for a snapshot of what lay beneath the streets of New York City in the 1950s:

> There are 7,000 miles of gas mains, 5,000 miles of sewers, 2,200 miles of TV cables and 15,000,000 miles of telephone wires – enough to circle this planet six hundred times. Subway trains, 8,700 of them a day, thunder along 726 miles of track. There are 87 miles of high pressure steam lines for tasks as varied as pressing trousers and heating skyscrapers, steam pressured to 150 pounds per square inch, cooked to 1,000° Fahrenheit and driven through concrete reinforced pipes at nearly three hundred miles an hour. Some 19,000 miles of electrical cables bristle with current boosted to 69,000 volts (the normal wall outlet is 120 volts). Ordinary water courses through 5,528 miles of mains under such intense pressure that a single leak could cause the loss of 3,000,000 gallons a day.[26]

Occasionally the underground city pushes itself into view, revealing something of the precarious nature of our seemingly calm and ordered lives on the surface. Aside from the giant alligators of urban legend, among the many hazards that lurked beneath the surface it was not unusual for sewer gas explosions to blow off some of the many thousands of heavy drain covers, thereby 'sending grenade-like fragments in all directions'.[27] But familiar though such occurrences may be, Daley added, they are still infrequent enough that no one any longer marvels at the city's subterranean roots, simply because they exist for the most part unseen and thereby avoid becoming the object of concern. Yet this concealment can easily lead to the absence of a crucial realization:

that without the existence of such underworlds and the work of those who build, maintain and exercise vigilance over them, the city would fall into disorder.[28]

Because we have so effectively banished this underground world from the concerns that carry us forward in life as we go about our business above ground, it is only when the kind of obliviousness this produces is interrupted that we may, for a time, glimpse an alternate, suppressed network of causes within which we will always be caught, and which exposes the illusion that – for instance – we might occupy a world free of waste. As Bruno Latour, the so-called 'Prince of Networks' and theorist of our entanglement in processes and relations, both human-made and natural, would say: as human agents, we discover the fact that we are merely one of many 'actants' (entities that act and are acted upon) that together constitute a complex social reality.[29]

Today, London's 'super sewer', a 25-kilometre-long (15½ mi.) replacement for the Victorian system designed by Joseph Bazalgette, which aims to prevent waste leaking into the Thames, excites a great deal of interest among residents of the city. The disruption caused by its construction – which involves the boring of giant holes beneath the surface of the city – draws attention, much like the blockages caused by fatbergs and other forms of disruption to the system do, back to the filthy reality of human life.

To enter the spaces of these urban underworlds is, in effect, to move between what might seem like separate worlds. In truth they form part of a single continuity, one half of which remains obscure to us. Symbolically, though, the separation of the visible world from that which is concealed means that any descent into the sewers can easily seem analogous to the Virgilian epic voyage to the place of the dead, an underworld that itself was characterized as 'a wasteland of deepest night'.[30] It is only possible to make such a link between our own reality and the ancient myth because of the success with which societies have separated themselves – out of necessity, and precisely to avoid fatal disease and death – from the realm of what we might think of as the non-human human: the wastes and wastelands, which shift and move and seem to take on a life of their own beneath our cities.

The Sanitary Philosophy

In St Michael's Church, St Albans, there can be found a monument to Francis Bacon, author of works that sought to explore the bounds of human understanding. It was erected by his executor Thomas Meautys, we learn, in order to 'keep alive the memory of so great a man' – a man whose fame had seen numerous titles and accolades heaped upon him (Baron Verulam, Viscount St Albans, the Light of Sciences) but who was, no less than other mortals, subject to the same law of cosmic return. As his monument states, he died 'after he had unravelled all the secrets of Polity and the Natural Sciences', at which point he 'fulfilled the law of nature that all compounds are subject to dissolution'.

The lesson is clear: from the earth you came, and to the earth you return. The human body, like any other organic entity, is finally caught up in a process of material dissolution, breaking apart and so being returned to the universal condition that is symbolized by base matter. 'In death my dust will mingle with sticky, slimy substances in the moist compost,' writes the philosopher Michel Serres; 'this is where the limit lies: smells of life, beforehand; funereal fragrances beyond this threshold.'[31] In nineteenth-century London, as Christopher Hamlin has revealed, this fact was intimated wherever there was no longer space to contain the dead. If death represented a reduction to base matter in the return of the body to the earth, it was a condition likewise suggested in the sight of 'stagnant sewers and cesspools, in heaps of garbage and excrement, in churchyards so packed with bodies that corpse parts continually surfaced'.[32] As Mumford wrote in a sweeping view of what had overcome the industrial cities of a fast-modernizing world, the situation seemed to portend a reversal of living conditions:

> Lacking the first elements of cleanliness, lacking even a water supply, lacking sanitary regulations of any kind, lacking the open spaces and gardens of the early medieval city, which made cruder forms of sewage disposal possible, the new industrial towns became breeding places for disease.[33]

As such, the great metropolis drew the interest of an array of investigators, all of whom gave substance to the idea that it had become 'the locus of fear, disgust and fascination'.[34] And it is through the exploration of the nineteenth-century city as a site of disorder unlike anything previously known that waste – in one form or another – emerges as central to the thinking of a variety of social reformers, sanitarians and medical health officers.

Edwin Chadwick, whose landmark Parliamentary report on sanitary conditions among the British poor was published in 1842, linked disease with a filthy urban environment, stating that across the nation the poor were trapped by 'atmospheric impurities produced by decomposing animal and vegetable substances, by damp and filth, and close and overcrowded dwellings'.[35] The writings of others, such as the British journalists John Hollingshead and Henry Mayhew, would echo such sentiments, although when it came to waste, they each had their own specific concerns.

Chadwick, looking to combat the dirty living conditions that were causes of illness and death, proposed a range of sanitary interventions to improve the quality of water. He advocated the construction of new sewers and the cleaning of streets and recommended greater ventilation within and between homes and dwellings to combat conditions in the worst slums, where families were often crammed into confined alleys that were dark and dirty refuges for a variety of ills. But more than just providing a set of recommendations, what Chadwick had produced, the historian Mary Poovey later wrote, was a new way of seeing: an 'image' of the city as a 'social body', which brought into focus the need for the government to make a 'medical intervention' to treat that body.[36] One decision made in the wake of Chadwick's report was to allow the Thames to be used to carry away the city's sewage, as it had done in earlier times.[37] This was a development that meant the choking city, like some constipated patient, was able to free itself from its excremental burden, which was now released into the river – only to turn 'its once sparkling waters into one big open sewer, brown and sluggish with its unwanted cargo'.[38] It was becoming a common view that 'London had gorged itself and become obese,' its 'main arteries . . . becoming choked'.[39]

The problem of the disposal of sewage was, for many nineteenth-century advocates of returning excremental waste to the land, characterized by practices that contradicted 'a sort of cosmic sanitary dualism'. The failure to return the plant nutrients taken from the land by humans to be consumed as food back to the land as manure – as had been the case in medieval towns and cities, with their gardens and fields beyond – was only one source of 'humanity's perversion of natural cycles'.[40] Modern sanitation, in seeking to banish the foul matter that resulted from human life under conditions of urban growth, accelerated and thus disrupted practices that had been in more or less harmonious relation to nature since time immemorial. Advocates of this sanitary dualism wrote encomiums to remind the public that the entirety of civilization depended on the products of agriculture, 'which in turn depended on restoring to the land the fertilizing elements taken from it'.[41] The only solution was to allow excremental matter to decompose in its own time 'into the simple inorganic forms plants utilized' and then return it to the land.[42] Otherwise, it would itself be wasted, which is to say, flushed out to sea.[43]

In 1855 one eminent British Victorian, the scientist Michael Faraday, wrote to *The Times* complaining of the state of the Thames, noting that 'the whole of the river was an opaque brown fluid.'[44] In true scientific fashion he devised an experiment to test whether or not his senses might have been deceiving him as to its opacity, and then carried it out at many points along the river, as he reported in the same letter:

> I tore up some white cards into pieces, moistened them so as to make them sink easily below the surface, and then dropped some of these pieces into the water . . . before they had sunk an inch below the surface they were indistinguishable though the sun shone brightly at the time; and when the pieces fell edgeways the lower part was hidden from sight before the upper part was under water. The smell was very bad, and common to the whole of the water; it was the same as that which now comes up from the gully-holes in the streets; the whole river was for the time a real sewer.[45]

These are sentiments that recall the opening passage of Charles Dickens's *Our Mutual Friend* (1865), when the reader is introduced to Gaffer Hexam and his daughter Lizzie as they tow a corpse back to shore, their vessel 'allied to the bottom of the river rather than the surface, by reason of the slime and ooze with which it was covered'.[46] The symbolism of the Thames as a murky body of water that had become as darkened as the befogged streets – and its banks at low tide revealing the presence of rubbish and rotting wastes – was great; not only because the river was symbolically central to a belief in the city's vitality but because it functioned effectively as the city's main sewer until the arrival of Joseph Bazalgette's new system for diverting London's wastes into a series of new underground tunnels, completed between 1865 and 1875.[47] For Henry Mayhew, whose gargantuan four-volume

This illustration, from an 1850 issue of *Punch* magazine, reflects the state of the Thames prior to Joseph Bazalgette's new sewer system.

study *London Labour and the London Poor* was published just prior to this (three volumes in 1851 and a fourth in 1861), this meant that if the tide was not adequately carrying this sewage out to sea, then the people of the city were likely drinking 'a solution of our own faeces'.[48] But what flowed into sewers or out to sea was only one facet of a phenomenon that took the city in its grip, creating blockages of waste wherever one looked.

We stand both inside and apart from nature. Material wastes, from excrement to rubbish, are constituted not just from the remainders of what we take and consume of nature in various ways, but more specifically from the difficulty, unwillingness or impossibility of it being reintegrated into nature. This is not a small problem but rather something that gets to the essence of how human living actually manifests itself: it is both inside and outside natural processes, and it is this externality of life, the abstraction from nature, that produces things that are difficult to assimilate back into nature and natural processes.

Urban Accumulations

The archaeological remains that helped to identify the site of the ancient city of Troy reveal something of the material traces of this civilization. Now compressed and layered, the leftovers of this long-buried past can only be accessed by cutting into the ground, creating openings in the surface of the present, and once again into a space below – another kind of underworld. Archaeology, indeed, rests on the belief that the evidence of previous times and places lies, wherever we may stand, beneath our feet.

The world we live in today – especially those places that have long been inhabited – is built and has risen imperceptibly over long periods of time on top of the rubbish of prior eras, which has served as the foundations, the ballast, of new worlds. The first human settlements, writes Robert Pogue Harrison in *The Dominion of the Dead* (2003), were those sites created for the burial of the dead, established 'so that their legacies could be retrieved, and their after-lives perpetuated'.[49] The living maintained their connection to the dead by digging into the ground to make space for those now departed but in need of a resting place. The burial of waste stuff at the site of Troy, although the result of practices that subsequently left behind valuable archaeological remains, was in no way as intentional as the burial of the dead; rather, it was more to do with the absence of organized waste management. Nonetheless, what it reveals more generally is that such remains are no less revealing of who we are and why we are where we are, than those sites that were established to protect the burial places of our ancestors.

With reference to periods before written sources exist, we can only rely on what little evidence there is to gain any kind of picture of the role of material wastes in everyday life. Even when written records came into use, and when so much of life went unrecorded, we might wonder who would care to keep details of the various kinds of rubbish or analyse the phenomenon of waste. There was little enough detail of the world of valuable things recorded. There is, as a result, a great deal that can only be inferred from archaeological evidence. Excavations at the site of ancient Troy, in modern-day Turkey, which uncovered several distinct city formations dating to different eras separated by long periods of time, suggested that at the time of the formation known as Troy II (2500–2300 BC) everyday activities took place within dwellings in which the floors were littered with shells, bones, potsherds and other rubbish, which was allowed to accumulate until people were wading 'ankle deep in trash and garbage'.[50] Once the situation became intolerable, the floor was probably covered over with a new layer of clay. In such manner the new city formations of Troy slowly grew on top of old ones (just as in Rome, the rising ground

Early Bronze Age ritual objects recovered from the site of ancient Troy.

level aided the enclosure of the watercourse that became the first covered sewer). But in practical terms, it meant that, conversely, the city was gradually sinking into the ground. These kinds of disposal habits raised the level of the domestic interior and led eventually to diminishing headroom, a problem that could only be addressed at that point by either abandoning the dwelling and rebuilding elsewhere or raising the height of the roof. At Troy, and in other places where waste stuff was compacted into the ground, there would be much of value that was buried – not, of course, to those who covered it over, but rather to the archaeologists of the future. In the words of one field guide on these sites, the wastes of previous eras left behind 'beautiful stratigraphy'.[51]

Such insights aside, little is known or can be said with confidence about the reality of living with waste in the distant past. For instance, from the plays of Aristophanes we might conclude that men in ancient Greek cities defecated in the streets, although other evidence suggests that well-regulated homes employed potties and other vessels to contain bodily wastes, and that the contents of domestic cesspits had to be removed to places well beyond the bounds of the city.[52] More broadly, it is also probable that periodic epidemics in antiquity would have been caused by sanitary problems, not least the filth and rubbish piled in the streets of cities that likely had no organized refuse disposal, thus attracting flies, rats and mosquitoes. Such nuisances, or the threat of them, will arguably continue to exist wherever people are found settled in places.

Given the paucity of written sources, the effects of waste may be inferred more generally from the facts that are known about the growth of towns and cities, and the knowledge that it is almost always reactive measures that are brought to bear when it comes to the management of waste matter. The commonplace view of European civilization in the medieval period – particularly the early part – as one of general decline was, for a long time, widely acknowledged. While more generally this decline theory is now widely disputed, the practical effect of the collapse of the Roman Empire in the fifth century hastened changes in economic and social life, the disappearance of central power and a decline of learning; 'urban populations decreased dramatically and many towns vanished altogether,' noted the medieval historian Vito Fumagalli in his 1994 book *Landscapes of Fear*.[53] Without order or control, waste quickly accumulates wherever human life is to be found.

In some representations of the historical past, Roman London – with its spas, bath houses and latrines – may have stood as a pre-modern example of the order and cleanliness of urban life, but this should not, as Peter Ackroyd suggests, conceal the fact that the city was likely not the sparkling marble wonder that might be lodged in our minds (derived principally from twentieth-century cinema and television recreations). And that is simply because what is normally left out of such visions of the past, of course, is the valueless and the discarded, which in Roman London would have included its middens and rubbish heaps, which might have been largely made up of 'the bones of oxen, goats, pigs and horses'.[54] Anything that could be scavenged would have been a target for ravens, 'always ready to consume offending garbage littered upon the street'.[55]

As this implies, the streets were where the remains of the butchering and food trades – later described as 'garbage', a term that referred specifically to the parts of a slaughtered or dead animal – would often end up. What made up this category of remainder was generally the offal that was beyond acceptable use – though in the absence of better cuts of meat it may sometimes have been consumed. Such contrasting ends or uses provide another example of how 'waste', whatever form it may take, frequently refers to a category of shifting use or value that depends on context and circumstance: the desperate and hungry will eat what might otherwise be off-limits. Offal, indeed, is a term that can be seen to have mirrored certain moral attitudes to what was considered acceptable or repugnant as food for human consumption. It referred to meat unfit for consumption (putrid flesh or carrion), and assorted offcuts that were usually deemed fit only for scavenging birds and animals (head, tail, kidneys, heart, tongue, liver and so on), but also to parts of an animal that were still at other times and in different contexts prepared for human consumption.[56] The birds of prey that fed on this (ravens and kites) were not only valued and protected for the very reason that they were part of how waste was informally managed (when, in later periods, they would be hunted) but were so commonplace that one might have thought they had been domesticated – 'so tame that they would snatch a piece of bread and butter from a child's hands'.[57]

However, we have only a limited insight into what the reality was, simply because our knowledge rests on the patchiness of the historical record. Yet it is probably reasonable to assume that there would have existed waste disposal and management habits or practices that remained constant from one generation to the next, and over long periods of time.

Moving forward a thousand years, some of the most detailed analysis of the significance of waste in the late Middle Ages comes from the research of a historian writing in the 1930s, Ernest L. Sabine, and published in the journal *Speculum*. The articles in question are, admittedly, limited to the case of one place – London in the twelfth to fifteenth centuries – but they nonetheless represent one of the only sustained attempts to understand the phenomenon of waste in the premodern period. One motivating factor for Sabine's studies was the need he saw to dispel some of the misconceptions about the nature of life in London during that period in its history, which had tended to view the city as a place consumed by filth and disorder.[58] A flavour of this view is found in Ackroyd's historical snapshots of London as the site of mounting filth that would raise the level of its ground upon the refuse of previous centuries. The 'discarded and forgotten objects, left among old foundations', he writes, 'help to support the weight of the modern city'.[59] In such images of the city, it is a place where 'the streets bear the marks of waste':

> Maiden Lane is named after middens, Pudding Lane after the 'pudding' sent down to the dung boats moored on the Thames. Public dumps were also known as laystalls and there is still a Laystall Street in Clerkenwell. Sherborne Lane was once known as Shiteburn Lane.[60]

In the thirteenth century, London was situated principally on the north side of the Thames and had a population of around 14,000, most of whom resided within the walls of the city. If the streets, as Ackroyd writes, would later commemorate the place as one apparently overtaken by waste, it would seem to be significant for our understanding. In a place geographically bounded and walled in, which became home to the waste-making bodies, habits and practices of ever-greater numbers of humans and animals through the centuries, it seems logical that life in the city would have been a battle against filth and all sorts of unwanted leftovers that were multiplying in kind and expanding in volume. However, the evidence that Sabine found led him to argue that it would be wrong to just take the worst descriptions that have survived in the written record for the norm. There were clearly 'degrees of cleanliness or uncleanliness', he noted, depending on what part of the city one looked at.[61] In some parts, the dung that tended to pile up in the streets, causing blockages and creating a foul stench, was subject to control through fines and would sometimes be recorded – in cases where offenders were penalized – by parish or city authorities. On the other hand, some places within the city walls where horses and cattle were kept, which must have become the source of prodigious amounts of manure, have left behind no records, which suggests that the city was successful in deterring misconduct. As Sabine argues, it might even be inferred that the very absence of records suggests a well-functioning system of waste management that only recorded breaches of the law.[62]

Away from the streets and the problems of animal wastes, the domestic sphere, of course, was another source of refuse and bodily waste that would end up finding its way on to the streets. In parts of the city, houses were built over four storeys. People living in the upper parts of houses would often throw 'at least the liquid part of their filth out the windows', which wasn't such a problem if it was raining as it would be washed into the gutters that carried it away.[63] But sometimes manual assistance was required to clear these channels, using water taken from nearby wells that had been filled with rainwater.[64]

While these observations can be made from the records that existed, Sabine was also keen to point out that relying on the written historical record can obscure the fact that life carried on under what we would think of as normal conditions before there was anyone around recording either the habits of daily living or the methods used to compel citizens to look after the places where they lived. In other words, city cleaning had 'long been established, not by written law, but by ancient custom'.[65] That is not to devalue the written records, for what they do show are examples of material waste that the authorities compelled people to deal with or dispose of in addition to the excremental and other wastes that may have been washed away in the street. These former included rushes used on the bare floors of houses; building refuse and earth from excavations on building sites; the contents of 'cesspools, especially of latrines or privies'; dung accumulations from passing traffic; and other 'rubbish', including sawdust, woodchips and straw littered about in the lanes. All of these had to be raked up and taken to appointed dumping grounds outside the city walls.[66]

In London, city cleaning was organized by 'wardmotes', which were meetings of the residents, including officials, of the city's specific, named wards, or subdivisions (wardmotes still exist in the 25 wards of the City of London). They met to consider matters relating to the maintenance of civic life and the management of roads, pavements and other public spaces. Aldermen, who were the heads of wards, could, if they felt it necessary, 'hire men for certain city-cleaning duties'.[67] In a broader sense the aldermen should be seen in terms of how a system of local controls was formed, and was able to carry on its business alongside other offices that had the power of compulsion when it came to dealing with the varieties of waste matter. These other offices were represented by the mayor and the common city council, who existed above the level of wards – and who therefore planned and organized the city cleaning and had the power to enforce their orders and even seize the property of those who, for example, left obstacles or nuisances in the streets. They might order special investigations into particular 'cases of abuse, such as those caused by throwing refuse' into the rivers that ran through the city: the Walbrook, the Fleet and the Thames.[68] In addition to this, matters related to the cleaning of the city could draw the attention of the monarch, who was able to intervene if problems persisted. There were many employed in keeping the city clean, Sabine tells us, including scavengers (at times known as 'surveyors of the pavements', people who monitored streets and passageways to ensure they were kept free of rubbish), constables, collectors of tolls, rakers and beadles (the last minor parish functionaries).[69] 'The scavengers acted as overseers,' we learn, 'supervising the cleaning of the streets by the rakers, and enforcing the abatement of nuisances caused by filth.'[70]

This all paints a picture that adds more detail to the broader perception of the period as one of material disorder in cities that were growing faster than they could be managed. Other details might be gleaned from written proclamations forbidding anyone from throwing quantities of surplus materials – rubbish, earth, gravel or dung – into the rivers and the running ditches at certain places, suggesting that locations where water flowed or was deep enough to conceal dumping would be ideal for disposing of waste illicitly.

The possibility of the city being overburdened by accumulations of excremental waste is evident also from the suggestion that London Bridge – a 'living bridge' dating from the late twelfth century, on which many houses were erected, reaching an estimated 140 dwellings by the fourteenth century – was akin to a large and very busy public toilet situated over the Thames, whose waters flowed below like an open sewer, it was said, 'for wise men to go over and for fools to go under'.[71] In places, the Thames would continue to be polluted in this way over the centuries ahead. In fact, by the end of the fourteenth century, the city's butchers, Sabine reveals, were commanded to carry their offal by boat to the middle of the river, where the currents moved most powerfully (later, piers were built out for this purpose), to minimize contamination of the water. There it could be dumped and then carried away, helping to protect what was still a source of water used in households for the purposes of cooking and drinking.[72]

Prior to this strategy, butchers had been allowed to throw entrails and other so-called 'filth' into streams like the Fleet, from which, Sabine notes, such waste would have been swept into the Thames. He remarks: 'The citizens of London seem to have considered the arrangement satisfactory, for they made no complaint of any nuisance being created thereby, even during the appalling outbreak of plague in 1349.'[73] At other times, during what were plague years in London, which included 1349, 1361, 1369, 1379, 1382, 1390, 1391 and 1407, there appears to have been some connection between the desire to improve sanitary conditions (which, 1349 aside, included regulating the disposal of 'butcher's filth') and the

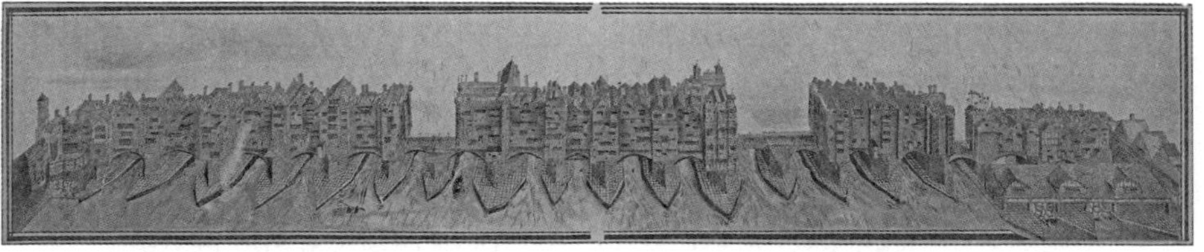

Old London Bridge, c. 1600, a time when it accommodated many houses, shown in a chromolithographed copy, 1881, by William Griggs of an older drawing.

fear of plague, as records show that such fears were the subject of 'agitation on the part of numerous citizens' in the latter half of the fourteenth century.[74]

The Leftovers

The landscape of modern Britain was changed irrevocably by the onset of deep coal mining, providing the source of energy that would power the first industrial revolution of the 1700s and 1800s. Starting in Britain and then developing elsewhere across Western Europe and beyond, the industrial world was a place darkened by smoke and soot, the atmosphere almost thickened by the prevalence of industry's grime and pollution, turning the built structures of cities black. People were sent underground to work in darkness in the pits that, once exhausted of their riches, were themselves termed 'wastes', and where miners moved around in subterranean tunnels. Above ground there was the constant din of machinery, the first intimation of a phenomenon – noise – that would itself eventually be described as a kind of pollutant. Nature was subordinated to human will, explored and appropriated for all it was worth. 'The atmosphere was being poisoned,' wrote environmental historian W. G. Hoskins, 'and in the Potteries and the Black Country especially, the landscape of Hell was foreshadowed.'[75] The north of England – soon to become known in near mythic terms as simply 'the North' – was transformed by skylines populated with giant chimney stacks, spouting the residues of their various productive activities into the vast skies, a scene that would spread throughout Europe and the rest of the world. The countless mines from which great resources were claimed created a hollowing-out that produced terrains of spoil heaps that persisted long into the twentieth century, and beyond in some places, by which time – and by virtue of outlasting the very industries that made them – they took on the appearance of almost alien, lunar-like landscapes. The industrial world of such places was indeed the world turned inside out and upside down, in which the demarcation between night and day was banished, and where a new underground world welcomed new inhabitants, whose occupations drew them into the bowels of the earth.

Mineral resources like coal, then being mined with a ferocity unknown to human history, constituted a form of energy bequeathed by geological processes and upheavals that had occurred many millions of years prior to human habitation of the landscapes where such resources were found. The coalfields of the North of England had been formed by the fossilized remains of swamp vegetation that had grown as thick, vast forest land hundreds of millions of years before. Here, as in other domains of human existence, life spread out and colonized the wilderness, or 'the waste', as it was also known, all the while creating more new remainders of its own. The sky, too, became a dump of sorts into which steam, smoke, dust, chemical fumes and dirt would circulate before falling onto and gripping the surfaces of the built environment. This fact became particularly evident in the polluted cities of a newly industrializing country. In Dickens's *Bleak House* (1853) the figure of the detective becomes, for the first time in the English novel, one way of making sense of the new and confusing world that had arisen, with the London fog – hovering, rolling, creeping, drooping and consuming the city's sky – serving as a 'metaphor for the obfuscations of social relations in the metropolis'.[76] In such accounts, the detective figure, as sociologist David Frisby argued, 'comforts city dwellers by suggesting that the city can be read and mastered, despite all appearances to the contrary'.[77]

The fog in Dickens is a character, a metaphor. But in reality the environment of London was also being transformed, if not its obfuscations directly tackled, by the expanding aims of a rational political will, expressed through a bureaucratic order that not only devised new and better means of draining the city's wastes but sought to bring more clarity to the social world, something exemplified in the collecting of statistics and the creation of new maps of the city. This all amounted to 'the consolidation of the objectification, abstraction and standardization of time and space' that would seek to dominate modern life.[78] Taking up an avowedly objective perspective on things, the statistical impulse was a somewhat different approach to controlling and comprehending the disorder of urban life from the one employed by those detective figures, whose knowledge of the city was perhaps more personal, more arcane: 'Statistics uncovered those laws of the social, knowledge of which would enable correct governance to take place. Social science itself grew up in intimate relation to this new way of thinking by means of number and social facts.'[79]

One figure who could be said to have been a forerunner of a new 'sociological' mode of investigation was Henry Mayhew, who in some ways indulged both of these investigative extremes. While his writings exemplified the attempt to enumerate the phenomenal extent of London's growth, he also embarked on gruesome expeditions into the darkest corners of the Dickensian city. The self-declared aim of Mayhew's *London Labour and the London Poor* was to detail the productive trades of the city. In fact, it was a work that exhaustively – and perhaps unintentionally – detailed the inescapable reality of London as a source of every kind of waste imaginable.[80]

Mayhew's interest was first captured by the unproductive activities of the poor, their existence outside the world of statistics, and embodied in the form of those such as the strange figures who could be found on the shores of the Thames at low tide – scavengers or mudlarks – prospecting for the most unlikely treasures that might be lodged in the sludge that made its way up to the banks of the river. The domain of waste that absorbs the poor, Simon Schaffer later noted with respect to Mayhew's work, is shown to be 'possessed of its own rationality, and thus, paradoxically, its own productivity'.[81] According to Mayhew, the mudlarks would 'scatter themselves along the shore, separating from each other', only to vanish from sight eventually as spectral figures moving 'among the craft lying about in every direction'.[82] Here, with the appropriate attention to the shifting surface and tides, they would find the stuff that refused to disappear, temporarily concealed by the muddy banks and the opaque body of the river.

In the economy of poverty as revealed by Mayhew, all objects and materials, no matter how degraded or used, retained some kind of value. 'Coals, bits of old iron, rope, bones, and copper nails', or any of a world of objects that might have fallen from the ships and boats on the Thames, or found their way into the river any other way, were sought out.[83] Perhaps these would be items that had slipped out of the pockets of floating corpses, which themselves – and aside from any valuables they may have had concealed upon them – could provide the means of making a living if one had a fishing vessel to use, as they could be robbed of their valuables before the tides went out to reveal the items to the mudlarks. In Dickens's *Our Mutual Friend*, as previously noted, we see the value of bodies fished out of the Thames as Lizzie Hexam and her father search the river for corpses and any valuables that might have been sunk with them.[84] For the mudlark, however, the only way to land a haul of the dubious river treasure was to wade through the silt, often until it was waist high. The sight of these abject figures – 'from mere childhood to positive decrepitude, crawling among the barges at the various wharfs along the river', as Mayhew wrote – was one of utter abandonment before the simple need to survive:

> it cannot be said that they are clad in rags, for they are scarcely half covered by the tattered indescribable things that serve them for clothing; their bodies are grimed with the foul soil of the river, and their torn garments stiffened up like boards with dirt of every possible description.[85]

From the muddy banks of the Thames at low tide to the 'foul labyrinths' of London's sewer tunnels, and through the streets and marketplaces frequented by a baffling variety of refuse-sellers and other street people, Mayhew collected facts, stories and impressions that filled out the pages of his four-volume work with the incontrovertible truth that, while waste stuff of all kinds may have represented the foul, disgusting and valueless remainders of productive society, the stuff did not disappear.[86] Rather, it merely found new life in the places that were overlooked and avoided by most. In the sewers he found men of a more enterprising nature and robust constitution than the helpless scavengers who waded on the banks of the Thames, many of whom were children and old women. Those prospecting in the sewers, by contrast, seemed capable of making a 'fair living' by finding 'great quantities of money' and were able to 'look down with a species of aristocratic contempt on the puny efforts of their less fortunate brethren the mudlarks'.[87]

While largely obscured from the bourgeois gaze and its environs, the sewermen could be seen near the river at the openings of sewer tunnels, wearing their 'long greasy velveteen coats, furnished with pockets of vast capacity' to hold what treasures they found, 'their nether limbs encased in dirty canvas trousers'.[88] Yet, with their proximity to the excremental tides that moved through the tunnels, there were certain dangers. The miasmic stench carried an intimation of death, whether that was a threat that was more imaginary than real. A careless explorer might easily sink into some unseen quagmire or lose their footing and end up buried under the decaying brickwork of a collapsing tunnel wall that had rotted away, Mayhew mused, 'through the continual action of the putrefying matter and moisture' that it transported. But even in the most dangerous sections of tunnel – those parts of the sewers crumbling away – 'clusters of articles' could still prove alluring enough to these determined prospectors, because almost everything was either of potential further use or suitable to be exchanged: 'iron, nails, and various scraps of metal, coins of every description'.[89] But it was a bounty accessible only to the most fearless adventurers, principally due to the number of rats that lay in wait and which many thought to be a certain means of death:

> Stories are told of sewer hunters beset by myriads of enormous rats, and slaying thousands of them in their struggle for life, till at length the swarms of the savage things overpowered them, and in a few days afterwards their skeletons were discovered picked to the very bones.[90]

All of that is to say that in Mayhew's work we are presented with a world of ubiquitous wastes. The aura of filth is everywhere: on the streets and inside homes; in the air that people breathed; and in the dust that was blown around and thereafter carried everywhere on the clothes of those who gathered on streets and in workplaces, shops and homes. 'London', Mayhew observed, 'is a perfect dust mill.'[91] Cast about by the galloping traffic of cabs and omnibuses, or blown by winds onto rooftops or into gaps and crevices where it might rest for a while, the dust would cling to the city, only to be washed away when it rained and then to mix once again with the macadam surface of the roads, not much more than material for the next dust storm.[92]

The sensory attack was unavoidable: the streets, Mayhew said, smelled 'like a stable yard'.[93] This was an impression that would lead him to further investigations into the extent of road traffic in London, and his subsequent astonishing calculations of the volume of horse manure resulting from the ceaseless movement of life that animated a city apparently bursting at the seams.[94] After duly delving into the matter at great length – minute detail being a characteristic of his four volumes, which contain entire histories of subjects such as prostitution – Mayhew declared that some 20,000 horses worked in the city on a daily basis. What, he wondered, became of the 'vast amount of filth' these animals deposited on the streets?[95] What follows is typical Mayhew – a detailed account of the horse dung and cattle droppings 'voided' (which is to say, gone to waste) on the streets of the capital. And if the figures did not speak for themselves, Mayhew was happy to

'The Sewer-Hunter', from Henry Mayhew's *London Labour and the London Poor*, vol. II (1861).

quote a report by the Board of Health 'as a much better guide' on this issue 'than the matter of quantity': 'Much of the horse-dung dropped in the London streets, under ordinary circumstances... dries and is pulverized, and with the common soil is carried into houses as dust, and dirties clothes and furniture.'[96]

But it did not end there. What, Mayhew wondered, happened to all these horses when they had reached the point of exhaustion and had to be replaced? That, too, was another unaccounted-for leftover. It was a question that seemed to become more complicated the closer he looked at it. 'There is, on an average,' Mayhew wrote, as ever in definitive voice, '1,000 horses slaughtered every week in London, and this, at 2*l*. 10*s*. each animal, would make the value of the "dead horses" of the metropolis amount to 130,000*l* per annum.'[97]

Unsurprisingly, in an economy so dependent on horsepower, the city had developed ways and means of making use of the horse carcasses. Numerous trades depended on the reuse of a variety of equine-derived by-products. The bulk of the remaining carcass became feed for other animals; the skin was used in tanning, the hooves for glue; old shoes and nails were reduced and reused; the hair of the mane and tail went into the manufacture of hats, laces, brushes, sieves and other items; the bones were used in manure and the intestines by makers of gut (string) and also in manure; and the fat was melted down and used as cart grease and harness oil.

Not given easily to surprise after the discoveries that were reported in the first volume of his work, Mayhew was nonetheless impressed enough to remark in his second volume that the disposal of horses was a 'curious trade'.[98] Further afield in France, he noted, the disposal of horses was even more complicated: the best parts of the flesh were eaten by those who worked in the slaughterhouse, while other desperate characters would hang around waiting for parts of the carcass that otherwise was sold to feed animals.[99] His apparent admiration for the French method of transferring rotting carcass into use-value is completed by the gory revelation of how what then remained of putrid flesh of the dead animal is once again made to teem with life, and thence to produce food for other living creatures:

> A pile of pieces of flesh, six inches in height, layer on layer, is slightly covered with hay or straw; the flies soon deposit their eggs in the attractive matter, and thus maggots are bred, the most of which are used as food for pheasants, and in a smaller degree of domestic fowls, and as baits for fish. These maggots give, or are supposed to give, a 'game flavour' to poultry, and a very 'high' flavour to pheasants.[100]

Going even further, and drawing on the observations made in Paris by Charles Babbage in his *On the Economy of Machinery and Manufactures* (1832), it seems that horse parts were used to lure and trap that other menace of the filthy city, the rat, thus extending the productivity of apparent waste to even greater lengths.[101] As Mayhew noted:

> The carcass of a horse is placed in a room, into which the rats gain access through openings in the floor contrived for the purpose. At night the rats are lured by their keenness of scent to the room, and lured in numbers; the openings are then closed, and they are prisoners. In one room 16,000 were killed in four weeks.[102]

But, while an example such as the Paris one above offered a means to see how things worked in other places, the overwhelming impression obtained from *London Labour* is delivered through Mayhew's spectacularly detailed account of the wastes of the city of London. The filth that appeared to grow prodigiously, to multiply and expand, and to infuse and perfume the air with a baffling admixture of microscopic waste material, in the end suggested the possible collapse of human efforts to contain waste in the face of the temporal disorder it represented. Filth, dirt, dust and putrescent matter were all signs that time did not stand still; proof, indeed, that any lapse in the work and habits of cleaning and ordering that, it was thought, were the source of culture or civilization would see such varieties of waste return as a counter-agent of civilizational progress. Filth, in this case, represented all manner of monstrous matter, threatening to return human life to some universal natural condition, as if to intimate a reunion with the mulch or mud from which life first sprang.

Out of the Slum City

If the work of Mayhew involved plunging into the depths of the mutating city to try and make sense of it all, there would soon be other perspectives that both responded to the conditions he found and attempted to reimagine or design a future in which something akin to the balance between society and nature that the sanitarians sought might be achieved. In the utopian vision of Ebenezer Howard's 'Central City–Garden City' concept, as laid out in his *Garden Cities of To-Morrow* (1902), we see a model city that accommodates all, giving space even to the inebriates and the insane, who had previously been left to roam aimlessly if they were not locked up. And it was through the judicious ordering of space that this utopia would aim to attain a harmony that, when compared to the filthy industrial city and its slums, would reconnect people with nature.

The British suburbs, although a distinct development from such garden city plans, represented a version of this ideal, expressing in their own way that it might be possible to escape the filth and disorder of the city. In the twentieth century, the suburbs

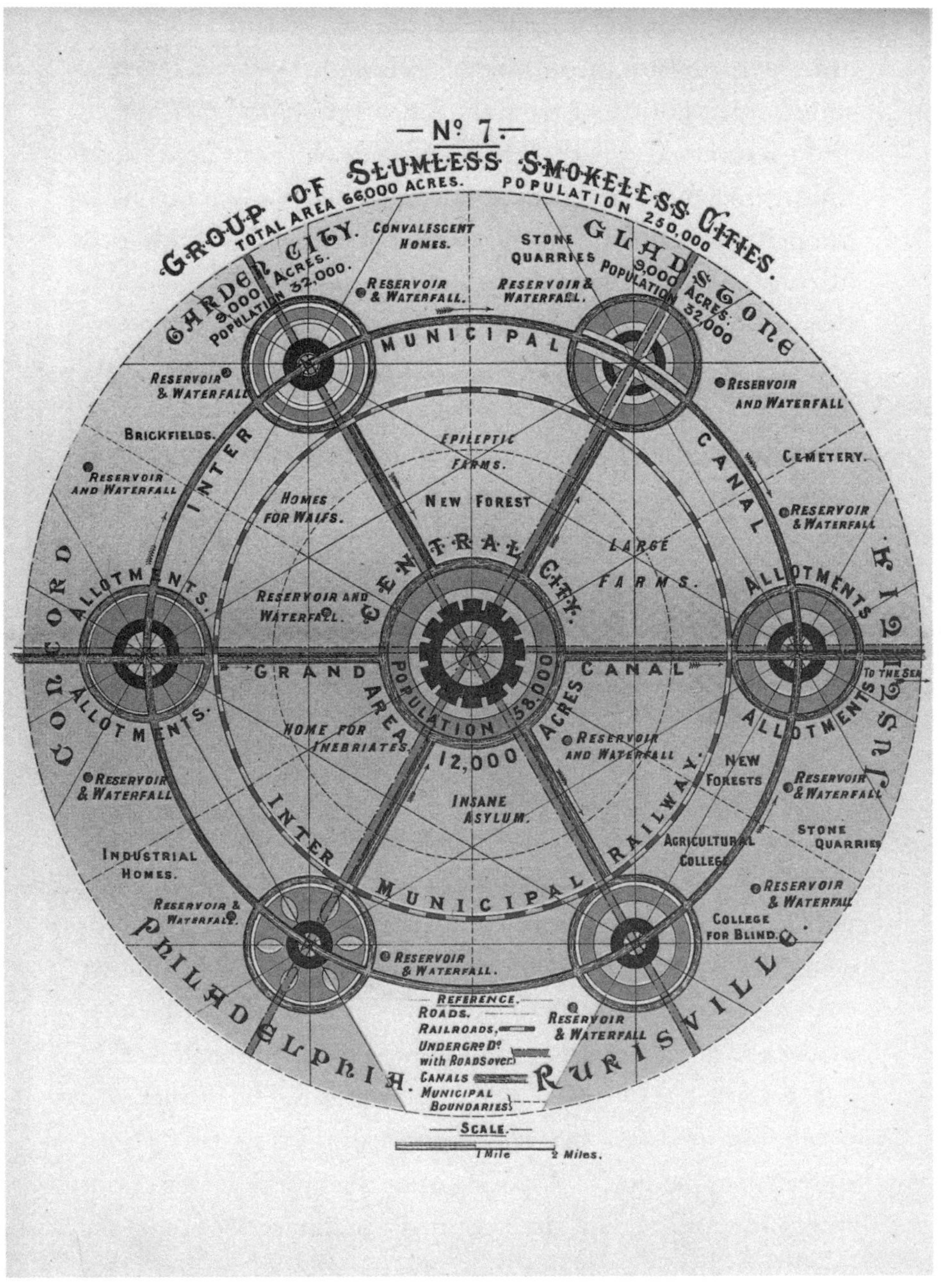

Ebenezer Howard's 'Central City–Garden City' diagram, from his *Garden Cities of To-Morrow* (1902), showing a design in which the balance between the human and the natural might be restored.

would also be crucial for understanding the nature and significance of material waste in a modern consumer society. The new suburbs that flourished from the nineteenth century developed as the response to the Victorian city: they represent life cleaned up, ordered, made beautiful, even. Yet suburbia – as it expanded through the twentieth century – went hand in hand with advances in consumer packaging and storage, something that became essential to our ability to live even further apart from the processes of production that create and deliver the goods that we consume. The movement from slum to suburb also charts the changing nature of waste matter – in addition to the need to manage dirt, dust and the putrescent there is soon the need to deal with rubbish that comes to consist more and more of packaging materials. Other developments in the design of living spaces in the mid-twentieth century, especially those modelled on the example of Le Corbusier's Unité d'habitation in Marseille, France, allowed for greater concentrations of people in a place, resulting also in more concentrated volumes of waste and new mechanisms of automatic disposal.

The logic of 'out of sight, out of mind' can be seen in the modernism of mid-twentieth-century domestic design, one of whose signature aims was to usher in a new, cleaner world for its residents, who were finally to be released from the slums of another century. London's iconic Barbican Estate – along with other modernist sites (including those allowed to fall into ruin within a generation, like Quarry Hill in Leeds and Park Hill in Sheffield) – came outfitted with the Garchey waste disposal system ('a clean and efficient method of disposing of kitchen waste as well as tins and bottles of a certain size', claimed a brochure, 'reducing the smell and vermin problems'), making the disposal of items that would otherwise have needed to be carried to an outside bin easier. This was done by removing the stainless steel grid through which water from the sink was drained, checking that the object to be disposed of through the outlet pipe would be able to pass through a U-bend (an instruction card against which the user could measure objects was supplied) and then flushing with tap water. Items that would be too large to pass through the pipe, such as long or large bottles, were to be smashed up first before disposal by these means.[103]

But at a societal level, the obscurity to most of us of even the most elementary methods of waste management often makes the category of material waste, as a repository of disaggregated matter, something that might hold or conceal unwelcome surprises. These have included terrorist bombs; unwanted personal effects (I am reminded of the writer William Burroughs once telling a story of wanting to tear up his adolescent writings into little pieces and throw them into *someone else's* trash); discarded CT scans purporting to show Bob Dylan's cranium (sought after by mad fans who intend to go where no one else has ventured); unwanted children (a phenomenon in ancient Rome and modern Italy); and – of course – that fixture of Hollywood movies, the dead body in the dumpster.[104] Such examples reveal something of the belief, perhaps held unconsciously, that what is to be gotten rid of via the waste stream is going to a place that has no connection to the person who is doing the throwing away.

2
Objects: Consume, Accumulate, Destroy

From Oswald Spengler's *The Decline of the West* (1918–22) – which posits the idea that all cultures eventually decay – to the image of history as destruction in the writings of Walter Benjamin and in landmark modernist works such as T. S. Eliot's *The Waste Land* (1922), there emerges an idea of the inherent power possessed by fragments, ruins, relics and other orphaned wastes.

The Living Dead

The power of these wastes in such writings is to reveal some inherent truth about life, and thereby establish an understanding of how the essence of human existence lies within the dialectics of civilizational progress and decline. The possibility of a new consciousness of time and progress lay behind Benjamin's idiosyncratic notion of 'natural history'. The primary example of how he explicated this natural history is found in his now well-known *Arcades Project*. The remainders of this work – even in the various posthumously published forms – have been passed on to posterity as what is, in itself, a strange literary waste-book, consisting of hundreds of textual fragments from secondary sources and observations that relate to the experience of the world of the nineteenth-century Parisian arcades, represented in and through quotations, aphorisms and photographs (some that were added to Benjamin's papers to create the 1999 publication), and his own writing. This waste-book, as it has been handed down to us, creates a new form of literary montage that was ostensibly inspired by the ideas of Surrealism (but was perhaps also, and more simply, just in tune with the material reality of its own time). The use of montage in visual art had itself, by the time Benjamin was gathering his materials on nineteenth-century Paris, already been identified as a product of a specific modern sensibility, one that had resulted from a consciousness that had been subjected to assault, if not entirely reorientated, by the ravages of modern society with all its upheavals. The result was a fractured perspective that would become manifested in artistic practices of the second half of the twentieth century; works that – like Benjamin's writing – in some way matched up to a reality that seemed to be always slipping away into chaos and confusion, but was now able to accommodate itself, as Ernst Bloch wrote in 1935, 'in the ruins' of the times.[1]

Among the fragments in the published version of Benjamin's *Arcades Project*, one finds many snapshots hinting at time's relentless pursuit of the human, but none so kaleidoscopic as his consideration of one amateur historian whose acquaintance with the world of waste could be compared with that of Henry Mayhew. Maxime Du Camp was a writer and urban explorer of sorts, but, like some undercover prototype of the participant observer of the next century, he was also unafraid of undertaking work that many would have considered too lowly in status, gathering the evidence and experience that he fed into a six-volume study of the city of Paris published between 1869 and 1875. He worked as an 'omnibus conductor, street sweeper and sewerman', Benjamin tells us.[2] But Du Camp began having trouble with his eyesight, which led him to the realization that the process of ageing would soon diminish his powers of perception. As he wandered around the city one day, waiting for an optician to make him a new pair of spectacles, it suddenly occurred to him – to this 'sojourner through mute and dreary wastes where the sand consists of the dust of the dead' – to imagine a day when Paris, 'whose enormous breath now filled his senses, would itself be dead, as so many capitals of so many empires were dead'.[3] This suggested to him that a final work demanding completion would be a consideration of Paris as a ruin in waiting.[4]

In such an endeavour, had he carried it out, Du Camp would have been a forerunner of Benjamin, whose preference for the world of 'small and shabby objects' led him not into the Paris sewers – not quite – but rather to the well-stocked shelves of the Bibliothèque nationale. There, he consulted the bulging archives that would later be the star of Alain Resnais' 1956 film *Toute la mémoire du monde* (All the Memory in the World) ('To avoid bursting, it [the library] is continually burrowing underground,' the narrator tells us). Among everything else – the endless lists of examples and ephemera and events that he compiled in document folders – Benjamin was captivated by a different kind of dust: the 'dust of centuries', as volumes were taken from shelves and opened up to the beams of light that entered through the ceiling, exciting dancing motes of times now past and his own perspicacious mind.[5] Where this led Benjamin was to a concern to record the nature of everyday life, and to see in the everyday something of the nature of a modernity that he understood in Baudelairean terms as the consciousness of a fleeting and ever-changing present. In Benjamin's work, as Ben Highmore observes, 'everyday life registers the process of modernization as an incessant accumulation of debris.'[6]

One category of the leftover and unwanted that Benjamin was intensely curious about consisted of bourgeois artefacts, the kinds of objects or souvenirs which, as excess or leftover goods from recent eras, were to be found in the covered shopping arcades of Paris. By the time Benjamin became interested in them, a century after their apogee, these places stood as relics of a way of life that belonged to the previous century. When they first came into

The interior of a Paris arcade, 2018.

existence in the 1820s and '30s, Benjamin notes, they were spaces where the 'commerce in luxury items' took place.[7] It was in these narrow, covered passages, which Benjamin characterized as cave-like or almost underground worlds (the roofs, constructed of iron and glass, let in shafts of light from above), that one could find the 'fossils of an ur-animal presumed extinct: the consumers from the pre-imperial epoch of capitalism, the last dinosaurs of Europe'.[8] There seemed to be no logic or design, no order, to the way the goods were arrayed (aside from those in the junk shop):

> A world of secret affinities opens up within: palm tree and feather duster, hairdryer and Venus de Milo, prostheses and letter-writing manuals. The odalisque lies in wait next to the inkwell, and priestesses raise high the vessels into which we drop cigarette butts as incense offerings. These items on display are a rebus: how one ought to read here the birdseed in the fixative-pan, the flower seeds beside the binoculars, the broken screw atop the musical score, and the revolver above the goldfish bowl.[9]

By the late 1920s, when Benjamin was living and writing in Paris, the husks of these now faded miniature worlds confined within corridors of the arcades would become the haunts of one of the most important figures in his work, the *flâneur*, who in these glass-covered passages could find the space to wander idly, gazing into boutique displays in search of intoxication, the grounds for which were, Benjamin suggests, prepared by literary depictions of the city.[10] Of note among such works were Surrealist texts such as Louis Aragon's *Paris Peasant* (1926) and André Breton's *Nadja* (1928), flights of the imagination in which 'the past is revealed to us in the present' and which reconfigure reality through 'dream-worlds of the city'.[11] Such literature functioned, in other words, as a narcotic trigger, a necessary preparation for ventures into an urban landscape that was now remade as a 'backdrop for the dreaming idler': 'The study of these books constituted a second existence, already wholly predisposed toward dreaming; and what the flâneur learned from them took form and figure during an afternoon walk before the apéritif.'[12]

Benjamin, though, was first drawn to the arcades for more material reasons. The iron construction of their coverings signalled something new to him, insofar as the use of iron permitted a new architecture that seemed designed for temporary purposes, as was the case with the Crystal Palace, designed for the Great Exhibition of 1851 in London, with 'its vaults of glittering glass and nearly invisible iron tracery', which for many observers made the industrial marvels contained in the displays it housed pale in comparison to the spectacle of glass and iron itself.[13] In this way, Benjamin saw iron as one of the principal conditions for the creation of the Paris arcades as revolutionary ambient spaces: new centres for 'commerce in luxury items' and all the other cultural associations that would flow from them in their nineteenth-century 'high' phase of existence.[14] Quoting from a guide to the city published soon after the arcades came into being in the 1820s, he highlights the view that 'the *passage* [arcade] is a city, a world in miniature.'[15] Later, in the period of their demise, they often became spaces that contained an abundance of junk.

As a construction material, iron would be crucial in the building of numerous modern spaces of transit and exchange (train stations, bridges, department stores), especially suited to newly changing urban landscapes of the era – and thus giving the edifices that those prefabricated iron beams supported an air of impermanence, as if the ease of construction guaranteed a corresponding ease of dismantling and, therefore, a kind of obsolescence in waiting. Nonetheless, the 'Empire', as Benjamin notes (referring to the Second Empire of Napoleon III, 1852–70), saw in iron construction techniques a means of quickly erecting a new world that might revive something that seemed to hint at the eternal and immutable, 'architecture in the classical Greek sense', which might serve as a means of projecting imperial ambitions.[16]

Just as the arcades were architectural novelties from an earlier time, so too the world that opened out inside their precincts appeared to belong to the past. This much was recounted by Benjamin in a text from 1929, one of several attempts to write a newspaper article on these spaces in collaboration with Franz Hessel. Together they entered a narrow, dark corridor to be met by some shops, including 'a discount bookstore, in which colorful tied-up bundles tell of all sorts of failure, and a shop selling only buttons':

> Here was the last refuge of those infant prodigies that saw the light of day at the time of the world exhibitions: the briefcase with interior lighting, the meter-long pocket knife, or the patented umbrella handle with built-in watch and revolver. And near the degenerate giant creatures, aborted and broken-down matter.[17]

And it was through the unwanted souvenirs and novelties that had become, by the twentieth century, the stuff of these arcades – persisting beyond a point in time at which their vital energies might still be extant – that Benjamin would see how the world of goods that were no longer desired, this degraded, overlooked domain of life that philosophical speculation might have deemed unworthy of attention, could provide a key to understanding the nature of modernity itself. To focus philosophical attention in this direction was probably inevitable for one who had been excluded (perhaps even self-excluded, due to his unwillingness to play the game) from the academy. Benjamin thus avoided the concerns of academic philosophy – the abstractions of logic or metaphysical speculations – because they bore little relation to the material reality of life. Theodor Adorno, an interlocutor who sometimes struggled with the directions Benjamin's work would take, wrote – perhaps in equal parts in admiration and exasperation – that 'small glass balls containing a landscape upon which snow fell' were exactly the kinds of material objects that enlivened the mind of his friend:

> The French word for still-life, *nature morte*, could be written above the portals of his philosophical dungeons. The Hegelian concept of 'second nature', as the reification of self-estranged human relations, and also the Marxian category of 'commodity fetishism' occupy key positions in Benjamin's work. He is driven not merely to awaken congealed life in petrified objects – as in allegory – but also to scrutinize living things so that they present themselves as being ancient, 'ur-historical' and abruptly release their significance.[18]

For Benjamin, we might say, obsolete commodities are like ruins: in a simple sense they are the remnants of a past that was once the future. But they were also much more. They served as a point of entry for phantasmagoric encounters that could produce a dizzying vertigo of 'bottomless depths', revealing that the modern notion of progress was still a myth.[19] It was the myth that forgets its own broken and discarded past. 'He was drawn to the petrified, frozen or obsolete elements of civilization,' Adorno wrote, 'to everything in it devoid of domestic vitality no less irresistibly than is the collector to fossiles [*sic*] or to the plant in the herbarium.'[20]

Benjamin had, of course, another purpose in mind behind what has been described as a form of 'trash aesthetics'.[21] In his second draft outline of the *Arcades Project* from 1939, 'Paris, Capital of the Nineteenth Century' (also known as the 'Exposé of 1939'), Benjamin begins by attacking the nineteenth-century conception of history that found expression through the elevation of the idea that classical civilization was the origin point for Western society. Benjamin's implicit aim was to confront the illusion 'expressed by Schopenhauer in the following formula: to seize the essence of history, it suffices to compare Herodotus and the morning newspaper'.[22] By the 1910s and '20s, as Benjamin was well aware, artists had already begun to deploy texts and images cut from newspapers and magazines on canvases, taking their materials from mass-media products that were typically thrown away and replaced with newer and more topical versions on a daily or weekly basis. The work on the arcades itself was described by Benjamin as a 'dialectical' fairy tale of sorts, an account of the past that would twist historical analysis around so that it was inclined more towards the temporality of the newspaper; or, in other terms, the *materiality* of present life, in which history manifestly did not exist before or beyond the consciousness of the observer (or indeed, the interpreter of the present) as some immutable inheritance of a civilization rooted thousands of years in the past.

In pursuit of an understanding of the everyday, Benjamin would therefore plunge with equal enthusiasm, as Esther Leslie notes, into the spaces where history might be encountered – archives – and the spaces of the everyday – the flea market. In such places Leslie describes Benjamin 'snatching up words and things that throbbed for him with unfulfilled and utopian energies from the past', revealing 'the weft of the past stiflingly entangled in the warp of the present'.[23]

Susan Buck-Morss, who – prior to the English publication (as *The Arcades Project*, 1999) of what Benjamin left behind of this work – undertook to reconstruct and attempted to 'write out' the *Arcades Project* from Benjamin's fragments and sketches, suggested that this work was intended to create a rather different kind of relationship to the past than that of history with a capital 'H'. It sought out, rather, a 'space of history' that set itself against what Benjamin saw as the myth of modernity's progress. This 'space of history' thus 'referred not only to the previous century, but to the ontogenetic, "natural" history of childhood – specifically the childhood of his own generation, born at the century's close'.[24] His generation, Benjamin felt, bore a special responsibility to '[reconstruct] the capacity for experience' that the trauma of modernity had left at the mercy of the dream world of commodities, which might challenge the idea of history-as-progress that reflected the dominant bourgeois ideology.[25]

But in addition to this, we must note that the form of the work itself, because Benjamin never completed it, will remain a literary waste-book; 'an astonishing ensemble of quotations and commentaries' whose final intended form will forever remain unknown.[26] In a way, of course, the published *Arcades Project* – running to over a thousand pages – emulates or presents a version of the almost ungraspable phenomenon that he was obsessed with. As already noted, the *Arcades Project* is a collection of multitudinous fragments that were organized into a series of 36 subjects.

These subjects reflected that, whatever it all had to do with history, this was not a history that aligned with the vertiginous narrative of modernity's origins in the distant past, but on the contrary one that explored how it was rooted in its own recent present (as reflected in sections that bore titles like 'Modes of Lighting', 'Idleness', 'Panorama', 'The Collector', 'Iron Construction', 'Boredom', 'Fashion', 'Conspiracies' and so on). Such content related to the experience of the world of the arcades in such a way as to create a literary montage that was ostensibly inspired by the idea that these fragments of the everyday would provide not only a glimpse into the Paris of the nineteenth century, but an accurate account of the history of the present. Benjamin thought that this material, when placed in the right combinations or juxtapositions ('construction out of facts. Construction with the complete elimination of theory', as he noted elliptically), would allow the dead world of modernity's past to come into its own, thus revealing its ceaseless cycles of destruction.[27] His position, as David Frisby noted, could be summed up in the phrase, 'I have nothing to say, only to show.'[28]

The Throwaway Society

In a time and place of bright lights, vacuum cleaners and kitchen-fitted waste-disposal systems ('no more waste – gone forever!'), rising affluence in post-war America heralded a consumer boom that would sweep away the old, quaint world within which people moved and goods had been circulated and exchanged. The consumer society had been predicted at the 1939–40 World's Fair held in New York, when General Motors' 'Futurama' exhibition placed the car at the centre of a society soon to arrive in which 'most people would be living in suburbs and commuting to work' on free-flowing highways.[29] As for the new people who would live in this future yet to come, Daniel Boorstin argues that a key aspect of the culture was the creation of new consumer communities, or what later would have been described in terms of consumer demographics. These were people amenable to mass marketing in a way that individual consumers had not been in the past. Such communities arose for a variety of reasons, but notably were created by new innovations in the promotion and sale of goods, the arrival of new kinds of shops and the forms of display they employed, and methods of mass production utilized to turn cheap materials into a world of constantly churning products, more affordable and more widely available than ever.[30] When Andy Warhol made the Coca-Cola bottle (and other products) icons of Pop art, this elevation of the everyday into art went hand-in-hand with his famous claim that no one, regardless of how rich or socially elevated their status, could buy a better Coke than the average person.

In the 1950s, liberated from wartime constraints on living, people in the United States embraced the new leisure opportunities that would henceforth consume their time, helping to ensure – when it came to the world of material goods – that a culture of repair was out of step with the new prosperity of the era. General Electric promoted the coming wonders of nuclear power, a development that was claimed to be the route to limitless free energy, suitable to power the new mass-produced home appliances that embodied the new age of leisure.[31] The trend was for manufacturers to develop and promote the use of more and more disposable goods as lifestyle options, to create the impression that it was fashionable for consumers to choose those new products that were 'made for convenience',

> with disposable cups, cutlery and nappies easier to replace than to clean, reducing the domestic workload. The range of materials it was considered acceptable to throw away started with paper and grew to include glass, ceramics, plastics, tin, aluminium and, increasingly with fast fashion, textiles.[32]

It is little wonder that, as all this was happening, books such as Vance Packard's *The Hidden Persuaders* (1957) and *The Waste Makers* (1960) and Alvin Toffler's *Future Shock* (1970) directed readers' attention to what seemed to be a newly accelerating reality, at the heart of which was the world of material goods. These goods ranged from cars to the smallest of kitchen appliances and food products and revealed the ways in which the new culture of speed had captured society. The books of Packard and Toffler, which were really contemporary reports concerned with the cultural landscape of the decades in which they were written and published, identified the importance of the system of artificially shortened product life cycles – the phenomenon termed 'planned obsolescence' – and why it had become of critical importance in maintaining an equilibrium of supply and demand, ensuring the system could continue in the same manner, without apparent end. Planned obsolescence ensured that products would be replaced by newer versions before they had exhausted their usefulness by stoking demand for new varieties of what might have been essentially the same thing. The centrality of obsolescence as an ideal ensured that the economy was focused more than it ever had been on developing products that were disposable, thereby giving birth to the idea of the throwaway society. The practices that we associate now with the idea of planned obsolescence originated more than half a century before Packard's and Toffler's books through the development of products like the Gillette disposable razor and other items that were intended – if possible – for single use, and 'then tossed in the trash'.[33]

The easy disposability this involved revolutionized the relationship that people had with the world of material objects and overturned a prior commercial tradition in which the idea of the durability of objects, materials and consumer products had been nurtured as a primary goal. That had been what, in the sale of goods and products, had established manufacturers, brands and retailers as trustworthy names in the marketplace. Emphasizing the longer lifespan of products had, of course, been a way of maintaining customer loyalty during economic conditions rather different to those that prevailed in the post-Second World War era, especially in times when saving was more important than spending. Back then, in the old way of thinking, the maxim 'waste not, want not' had meant something to people. But what was happening in the era of planned obsolescence was the consequence of a society that had surplus wealth to spend, and this phenonemon was now transforming itself through a broader acceleration across various facets of life that aimed to make it possible to squeeze in more of everything: more consumer goods and services, more leisure activities, more experiences and the creation of endless new needs that would also have to be satisfied in ways that could fit in with broader changes in society. It was all about letting go and being *carried away* (to quote the title of Rachel Bowlby's book on the history of shopping, which employs the double meaning of that phrase to refer to being whisked off one's feet by a succession of new products and experiences, as well as the carrying home of those goods).[34]

Taken together, these factors represent changes we still see occurring everywhere today. They were evident in the production and display of pre-packaged foods, which, since the mid-twentieth century, ensured that 'old' forms of food waste were either reduced or replaced in domestic rubbish bins by the packaging and containers that appeared with the rise of the supermarket as the focal point of food distribution. This was part of what has been termed 'the great clean-up', in which dirt and dust were increasingly banished from life, but only at the cost of new forms of trash.[35] The 'clean-up' brought us frozen foods, plastic containers as well as non-material elements of a new experience, which played out within the ambience of brightly lit environments that were stocked with an abundance of packaged goods.

At the time of its initial publication, Packard's *The Waste Makers* detailed the domains within which he saw waste-making as the real context of how consumers themselves were being remade – we might say 'reprogrammed' today – as new kinds of people: people who could be 'coaxed to indulge' in new forms of wastefulness by large corporations whose business was to widen the availability of disposable goods that would fit easily into the busier schedules of everyday life in the second half of the twentieth century.[36] The aim was to get people to leave the old behind and to embrace the new. Everyday life acquired a fresh look and feel. This was the 'Populuxe era', in the words of historian Thomas Hine, in which 'Americans were incessantly exhorted to look forward to the 21st century.'[37] Consider the rise of what became known as the TV dinner, a ready-made product designed for maximum convenience, requiring a minimum of effort on the part of the consumer. In its early days it was characterized by the disposable interior packaging that would carry the food from outer package into ovens and boiling pans of water: 'trays that can be cooked, bags that can be boiled, bowls and other eating utensils that can be discarded to eliminate dishwashing'.[38]

Noting that it was now possible to cook a steak and other foods and have no mess left over to clean up, Packard observed that this development was all a result of the ease with which packaging materials – including novelties such as throwaway cooking pans that came attached to steaks – could seemingly be made to vanish after a single use:

> Muffins come in throw-away baking tins. Hungarian goulash began being offered in throw-away plastic boil-in bags. Charles Mortimer, the chairman of General Foods, exulted that Mrs Consumer was learning to expect 'dinners which can be popped into the oven, heated, served in the pan, and the pan tossed into the trash can after dinner!'[39]

What was also new in the throwaway society was the nature of the stuff that was thrown away. In terms of domestic or household waste, it was no longer – as may have been the case half a century earlier – just food scraps, ashes or papers, but it would also, and significantly, include more synthetic plastics. These new materials were extremely pliable and able to be injected or moulded into any shape, something that allowed for the creation of more and more specially designed products – with different looks, shapes, sizes and colours – that all had plastic shells.

It is true to say that many goods had been made, as far back as we can look, from 'natural' plastics – if plastic here refers to pliable material that can be shaped to meet particular needs – including 'animal horn and hoof, tortoiseshell, bone, ivory, guttapercha, shellac, glue, and other compounds'.[40] But it took the arrival of synthetic plastics in the early twentieth century to make a real difference to methods of mass production and subsequently to how life was experienced through relationships with an expanding world of consumer goods.[41] It is now almost impossible to imagine life without synthetic plastics of all kinds, so ubiquitous have they become since the 1930s, when synthetic cellophane was first used as a protective material for foods.[42]

Plastic was used to develop a baffling array of new products, but it was also utilized in new forms of packaging that could at once protect or prolong the life of a product while also being designed to be instantly disposable. And because it was a material that could be produced in a wide variety of colours that transformed the way consumer objects looked, it contributed enormously to the production of consumer goods that relied on the idea that people would become bored with the same forms or colours, and so buy new versions of the same – a hand-held portable radio, for instance – in a new shape, size or colour. In short, plastic changed the way the world looked. The versatility of the material would also play an important part in how goods were able to be packaged, shipped and displayed in shops and used in the home. It made all of those things easier, but purely as a packaging material it would end up leaving only more physical matter to be disposed of. An example of one of the worst kinds of use is so-called 'blister' packaging, in which the plastic material is thermoformed to the shape of a product, to create astonishingly annoying extra layers or casings of plastic. It protects the product during packing in the factory and shipping out to shops, and facilitates racking and display, but at the same time it often requires the buyer to resort to sharp implements to break through this tough exterior to get to the actual product. Another example is the non-returnable and non-biodegradable plastic bottle or container, with which we are now so familiar, that first began to proliferate in the 1970s.[43] A material made for convenience and easy to dispose of, its lifespan may extend 'from 450 years to forever'.[44]

Plastic materials were capable of being made into a bewildering array of products, which touched every aspect of life in the 20th century.

Selves Sorted

As French philosopher Gaston Bachelard wrote in his *Poetics of Space* (1958), 'a house that has been experienced is not an inert box. Inhabited space transcends geometrical space.'[45] The possessions that we live with and through may establish the nature of domestic inhabitation, but they are also conduits between inside and outside – materially, spatially, psychologically – and, as such, demarcate the private and public realms. The meaning we attach to possessions indicates the fact that our sense of who we are is bound up with the extensions of self that end up turning abstract space into a meaningful place, as anyone who has moved into a new home – or indeed, been forced to live in some temporary abode and unable to set up home – will be aware.

The domestic space of the home is thus the physical interior of a shell that forms a connection to the psychological interior: the mind of the person. It may be difficult to conceive of the home as a dwelling without thinking about the role that objects and belongings, or the personal choices that inform the decor of the interior, play in it; or about what role the display of personal mementos plays in contributing to a larger idea of self beyond the significance of any single object taken on its own. These possessions surely help us to establish a sense of who we are that is both inwardly held and also projected outwards. As such, the interior spaces of the home might also be thought of in terms of the psychological interior. Possessions, for Hungarian American psychologist Mihaly Csikszentmihalyi, whose concept of 'flow' was popularized in the 1970s, formed part of the 'flow of psychic energy' in the domestic space, creating the familiar environment that is, at least in theory, under our command, and therefore allows us to channel 'psychic energy more effectively within it'.[46] In these terms, the contrasting act of removing possessions to a separate or intermediate space – an abstract space – indicates a change in how psychic

A dilapidated-looking warehouse with a new road-facing facade remade into storage space for those excess possessions, Manchester, UK, 2014.

energy is being used and directed. With a reduction in this psychic energy, Csikszentmihalyi suggests, we might say that the objects are halfway to being forgotten, whether or not that is the initial intention of moving them to another space. I prefer to think in simpler terms, of the inside, the outside and the in-between. The last is always potentially representative of how waste fluctuates between subjective values of negative and positive.

If it is true to say that the modern landfill site is akin to some hypertrophied growth to be found somewhere around the towns and cities in which we live, by the late twentieth century it would be joined in that respect by another by-product of consumer society: the ingeniously named 'self-storage' facility. Self-storage is a relatively new concept, certainly as it refers to a service that is as accessible as going to the supermarket. These giant, box-like facilities with their numerous cells (containers of various size and dimension), their availability 24/7, 365 days a year, and so on, started to sprout across the landscape of Britain about forty years ago. These are the places where one might spirit away some of the material excess of life without making a definitive act of dispossession, and without condemning personal belongings outright to instant death of the landfill kind, as might be the case when one runs out of time to find a more efficient or useful means of managing the objects' decline in one's affections. Belongings and possessions that once had a place on the inside but find themselves almost halfway towards the junk heap may be stored in these peculiar way stations, which seem to manifest the type of inert box that Bachelard contrasted with inhabited space.

The invention and growth of self-storage from the second half of the twentieth century reflects the many cultural changes that were set in motion as people came to live not just in a more acquisitive society, but in one that was also significantly more mobile. If home possessions said something about the person, then decluttering and depersonalizing – through the offloading of unnecessary or unloved objects – was a way of changing one's sense of self-identity. Andy Warhol, who in his declaration that the future would see everyone become famous for fifteen minutes, had his finger on the pulse of an accelerating society and may also have conceptualized self-storage before it was fully actualized as an aspect of life. In the collection of miscellaneous thoughts that was published as *The Philosophy of Andy Warhol* (1975), he raises the twin problems of having accumulated too much stuff and not having enough space to keep it. He remarks that, if it was not possible to live in 'one big empty space' or emulate Japanese minimalist living by rolling up and stowing away one's possessions, then the aim should be to have a very big closet, but ideally a separate space, detached, so as not to upset the mood of everyday life with concerns about the need to maintain a close attachment to its contents. 'If you live in New York,' he advised, 'your closet should be, at the very least, in New Jersey.'[47] That may have just been a dig at New Jersey's image as both the dumping ground for New York City's trash and leftovers and the place whose shores were actually polluted by the stuff that New York tipped into the waters along its coastline.[48]

First established in the United States in the 1960s, modern self-storage arrived in Britain in the 1980s, and in hindsight could be seen as evidence that we were already in the midst of a lifestyle revolution. Its ease of access supplemented and largely replaced older storage services that were intended for household goods packed up and left for longer periods of time, services that were directed towards a particular kind of client – perhaps the wealthy resident of more than one nation, with homes here and abroad – who needed space in which to keep their belongings until they could be shipped abroad or sent back to the location from which they had been temporarily moved in the first place. In those days, there was little idea that there could be a demand for storage space that, although located miles away and requiring a car to access, was almost as easily accessible as a garage or garden shed situated right beside the home.

Self-storage locations were also new, and had much to do with how the land around motorways and bypasses that skirted towns and cities could be utilized for commercial purposes; they formed a continuity with the warehouses, motorway intersections and traffic of a society that would come to be defined in terms of movement and, more broadly, a range of what social theorists termed 'mobilities'.[49] This much is implicit in the amusing name given to such a facility in Thomas Harris's 1988 novel *The Silence of the Lambs*, in which FBI agent in training Clarice Starling is sent looking for documents relating to a victim of the serial killer Hannibal Lecter. She discovers the location of the material she is after in a lock-up that Lecter had rented (payment in advance), almost ten years earlier, on the outskirts of Baltimore at 'Split City' storage. The location, we learn, is much the same as it would be in any city: on the fringes of things. The name 'Split City' refers, one assumes, not only to the ease with which people were able to leave their past lives behind but to the attractions of living a split existence, with all the excess and unwanted stuff of life enjoying its own secure space, out of sight, out of mind:

> Split City is a bleak place the wind blows through . . .
> a service industry to the mindless Brownian movement in our population; most of its business is storing the sundered chattels of divorce. Its units are stacked with living room suites, breakfast ensembles, spotted mattresses, toys, and the photographs of things that didn't work out.[50]

What Starling finds there after having to break into the unit ('Our customers value security,' she is told), of course, is not the documents she thinks she is chasing but rather something more gruesome: the severed head of one of Lecter's victims, preserved in a jar. The idea of self-storage was empty enough that the spaces it defined could be used for many purposes.

The other key feature of self-storage, aside from the security it offered, was 'self-access'; that is to say, the ability to enter or leave the storage space at one's leisure, 24 hours a day, all year round, making it eminently suitable in theory as a temporary repository for what one newspaper report, at the height of interest in these expanding facilities, referred to as 'day-to-day domestic overflow'.[51] But the idea didn't really begin to take off in Britain until the 1990s, and it may be no accident that this coincided with the era of lifestyle gurus urging people to simplify their lives through acts of dispossession.[52] Other reasons were floated to explain why self-storage was becoming, in both the UK and the USA, a necessary, although 'low-profile', service industry. One was that an ageing baby boomer generation was running out of space, as they faced down 'the caches of heirlooms and clutter in their parents' basements', as well as the problems of negotiating a variety of 'life events', not least of which was divorce:

> For the most part, storage units were meant to temporarily absorb the possessions of those in transition: moving, marrying or divorcing, or dealing with a death in the family. And the late twentieth century turned out to be a golden age of

life events in America, with peaking divorce rates and a rush of second- and third-home buying.[53]

In the UK, the self-storage boom era of the 1990s was also marked by the spectre of such places as repositories for more deadly kinds of cache. Self-storage was front-page news when the IRA, then in the process of carrying out a UK mainland bombing campaign, was discovered to have been using Abacus Self Storage in Hornsey, London, to store bomb-making paraphernalia.[54]

A sign of the times was a 1995 art installation that took place in another London self-storage facility, located, in the words of one contemporary report, in the 'unlovely London suburb of Wembley'.[55] *Self Storage* – as it was titled – was curated by record producer and one-time rock star Brian Eno in collaboration with the arts organization Artangel. By 1995, Eno had earned a reputation as someone willing to throw himself into any artistic endeavour that showed some potential, who just then happened to be a visiting professor at the Royal College of Art. What Eno was very good at was doing work that garnered media attention, and *Self Storage* – although taking place before the IRA bomb find – was still very topical.

As well as Eno's own contributions to the installation, the show featured the work of 21 staff and students whom he had rounded up from the Royal College of Art.[56] They were given access to a former tinfoil factory that had recently been transformed into a storage facility containing '650 self-storage units, some the size of a closet, others the size of a ballroom'.[57] The 26 units that would end up forming part of the show were prepared as multimedia enclosures, promising sights, sounds and smells for unsuspecting visitors, all apparently inspired by the idea of storage. In Eno's words, it was the nature and location of the space itself and the idea that 'you can just park bits of your life around the place' that grabbed his attention.[58] Eno had also recently graced the cover of the hip high-tech magazine *Wired*, where he declared that what made art important was that it didn't really do much beyond 'getting us used to the idea of enjoying uncertainty'.[59] And where *Self Storage* was concerned, uncertainty was surely the idea. What, in fact, lay behind all those locked doors? More severed heads? Bomb-making equipment? There was no knowing.

In a diary that Eno published the following year – titled *A Year with Swollen Appendices* – we see that much of the early months of 1995 were taken up with his preparations for the show and the regular and often fruitless visits to the storage facility. 'Why the fuck am I doing this?' he records after one particularly disappointing day on site.[60] 'At Wembley,' another entry reads, 'a dreadful, windswept place . . . lunch at the worst cafe on earth. Should be awarded a blue plaque.'[61] But as one of the reviews of *Self Storage* would later advise those who were thinking of attending the show, Eno may have been forgetting the fact that every self-storage outing involves a journey into the urban edgelands: 'arriving in this North London wasteland is a crucial part of the experience.'[62]

One of the aims of the show was to make the public curious about what exactly lay secured behind those hundreds of padlocked steel doors: it was, in the words of Artangel, like 'an industrial version of Hampton Court maze'.[63] Visitors were lured in by the voice of the show's other famous collaborator, Laurie Anderson, she of the voice as spectral instrument, which in this case had been subjected to the Eno treatment and filled the air as a very strange greeting: 'Hey hey hey yah hey hey yah hey I'm here . . . hey hey yah hey ah hey . . . yeah yeah here.'[64] This was one sound that led onto a pathway decorated with IKEA-style directional floor arrows, taking them from one lock-up to the next, where they'd make surprise discoveries – including the voice of Anderson again, who provided stories for each of the installations – or meet with some nightmarish soundscape devised by Eno himself (one sounded like a handful of razor blades being dragged over the surface of glass), perhaps apt for a location that was described as 'one hell of a creepy place'.[65]

Behind the doors that were opened to the public were a series of installations: objects such as the *Vizier of Memphis* (on loan from the British Museum and crammed into a closet-sized unit); a recreation of a crummy bedsit by artist Michael Callan; a woman (artist Michelle Griffiths) suspended in a tank of water fed by wires and oxygen tubes; and an enigmatic installation Eno had put together with three canine figures, a pyramid and some garden soil. 'I can't imagine a space like this existing at any time before the last ten years,' he told a reporter; and, in the same article: 'It is a funny modern idea to park your life out here, ten miles from London, and lead this Zen existence in town, coming out to pick things up when you need them.'[66]

Self Storage's location, just off London's North Circular Road, was the kind of place explored in the work of writer and psychogeographer Iain Sinclair; places at one with the giant home-furnishing retail sheds, supermarket-chain superstores and shopping malls that had come to occupy the outer rims of our cities as 'buffer zones between town and country'.[67] Such places, as the finance pages of British newspapers had started to note early in the 1980s, may have seemed empty or abandoned, but were full of something else: investment potential, something that would make them, in Sinclair's words, part of the 'distinguishing structure of future edgelands'.[68] The shells of abandoned factories and other structures with vast interiors that occupied temporary waste ground were especially suitable to be gutted and repurposed for self-storage. 'Property Conversion Increases Profitability', ran one advert in the business pages of a British national newspaper in 1990:

> Leslie Steel Storage Systems Ltd seek interested partners for business opportunities in Major European cities. We

> specialize in the refurbishment and conversion of old properties into self-storage centres ideal for short term rentals. The self-storage industry is now firmly established in the UK and offers excellent returns on investment.[69]

A decade later, on the cusp of the new millennium, and predicting that, whatever form regeneration was then taking in certain places, it was likely only a temporary measure, the journalist Tim Dowling would joke that the huge white elephant that had now taken over Greenwich Peninsula, the Millennium Dome, would probably only find its time and purpose in life twenty years hence, when it would make a perfect candidate to be transformed into deluxe self-storage units. 'Meridian Secure-Lok, London's premier self-storage facility,' he imagined some future promo literature boasting, 'introduces a special long-term storage rate for its regular customers.' 'Located in the former Millennium Dome. Secure-Lok is large enough to accommodate all storage needs, from family heirlooms to antique 747s. Secure-Lok is just five minutes from central London via the Blairway.'[70]

The 'Blairway' was a reference to British prime minister Tony Blair, whose government unveiled the Millennium Dome in 1999, and both Blair and the Dome, of course, feature as menacing and spectral presences in Sinclair's writings of the time. Speaking from a vantage point close to the Dome, one of the characters Sinclair encounters in his 2009 book *Hackney, That Rose-Red Empire* tells him that 'storage is the growth industry, twinned with the cult of minimalist lifestyles; empty apartments kept empty, as a long-term investment'.[71] A perfect choice for the well off yet precarious footballer, newly enriched by temporary promotion to the upper echelons of the game, and in London's East End perhaps bought as insurance, Sinclair muses, 'by West Ham players as a hedge against the annual threat of relegation' and the potential reversal of fortunes that such an eventuality might entail.[72]

Perhaps the strangest, and in many ways most tragic, example of self-storage's capacity to extend into abstract space is in the case of the posthumously famous photographer Vivian Maier. Maier spent a lifetime taking photographs of American street scenes while she worked as a nanny to earn a living, going unrecognized as an artist while she was alive. The many families she was employed by knew little more about her than the fact that, within the modest living space her role usually afforded her (often a single room), she liked to keep her private life to herself. Maier's negatives, mostly undeveloped and kept in a number of large cases and trunks, ended up in self-storage after she had retired as a nanny and ran out of space at home. Like a scene from one of the many TV shows that sprouted up in the 2000s – with titles like *Storage Wars*, *Container Wars*, *Auction Hunters* – about people who would make a career of sorts bidding on the unknown contents of storage units left by owners who had died, or by others forced, like Maier, to abandon their possessions, her life's work was only discovered after her storage provider auctioned her possessions because she was unable to keep up the payments to maintain use of the space. Maier, as is now well known through the films and newspaper stories that have appeared since the posthumous discovery of her photographs, not only 'carried her life around with her in boxes and suitcases' but kept everything, not least thousands of unprocessed rolls of film that today constitute an archive of her life as a photographer.[73] Her story perfectly illustrates the in-betweenness of the self-storage space: at once a provisional space to store junk or surplus valuables, and the source of unknown treasure. There is no telling what would have become of the countless rolls of undeveloped film if her trunks and suitcases and other belongings had not found their way to someone who had an interest in photography.

We might say that the temporality and ontological status of waste is to endure beyond need. But more than this, it is to be an object of consciousness. Yet whatever waste is, it shifts in and out of consciousness at the subjective level and has different meanings for those who stand in relation to it – which is why it may be constituted as something useful to one but of no value to another. This explains why we have flea markets and so-called dumpster divers, both of which illustrate how two contrasting value systems – those of the disposer and the scavenger – may meet, so to speak, on common ground.

Break Down

No one illustrated the nature of consumer acquisition and how deeply it had gone to the core of who we had become as a society as much as the English artist Michael Landy, in the 2001 performance *Break Down*. Much of Landy's work up to this point, which had frequently touched on issues of waste in various ways, seemed to be at odds with a new generation of British artists (dubbed the YBAs, for Young British Artists) with whom he was often lumped and who were the source of great media attention. The YBAs represented a phenomenon that seemed to share in the atmosphere of 'Cool Britannia', a label applied to a resurgent popular culture in Britain that saw the stars of pop music and the fashion world feted by politicians. Landy's works, such as *Christmas Tree, 1997* (a commercial wheelie bin overflowing with broken toys, a chewed-up teddy bear, the signs of excessive food and alcohol consumption, and, of course, a discarded Christmas tree), had come to the notice of the media partly because cleaners at one London gallery mistook some of his work for rubbish and threw it out.[74] *Break Down*, nonetheless, went somewhat further and took his interest in the hidden consequences of consumer society to what one might reasonably describe as an extreme of self-destruction insofar as it firmly located the self of late capitalist society in the acquisition of personal possessions. *Break Down* was a work that began with detailing every last thing that the artist owned. Singular in its setting and execution, it was a fourteen-day performance that took place in the most conspicuous of locations befitting its subject: a disused store on London's Oxford Street, home to flagship shops and department stores and a destination for thousands every day. 'Many of the objects to be shredded in *Break Down* were personal letters, photographs and memorabilia,' noted art critic Alex Coles:

> Other objects, even those with the most generic appearance, like a plastic lighter and a toilet brush, had been personalized in being gifted to Landy from a loved one or friend. Such is the extent to which the work incited the beholder into producing their own analysis of the artefacts on display.[75]

Break Down fed into contemporaneous public campaigns that sought to make people think twice before buying new goods (such as the 'Buy Nothing Day' campaign of 1999), but without forcing those who saw it into bouts of 'guilt-ridden self-questioning'.[76] One reason that Landy tried to avoid being preachy on this point was because he realized how intimately people associated their lives with their possessions, and how implicated most people were in consuming to excess. At the start of each day, the artist and his crew of dismantlers geared themselves up to the sound of David Bowie's 'Breaking Glass' as they donned hard hats and overalls.[77] Then, they proceeded to destroy everything Landy owned until it was all reduced to bags of ground-down matter of various kinds. This all took place in full public view – the glass storefront like a frame containing the performance – and under the gaze of passing shoppers who were likely in the vicinity precisely because they were exercising some consumer urge or other. The art of destruction was not new and, indeed, had been pursued as both subject and practice in a myriad of ways and through numerous media and approaches.[78] But Landy's art of destruction, because it was so intimately tied to to his own life as a consumer and was undertaken in recognition of how these possessions mediated his relationship to the world, was profound in its recognition of the fact that we all become what we consume and what we throw away. Once his possessions had been destroyed and the formerly recognizable forms – a car, an item of clothing, home appliances, all detailed in drawings and lists made by Landy – had been reduced to parts or ground into tiny fragments, there was no other use for it but to fill up a hole in the ground somewhere.[79]

This interest in waste had originated some years earlier when, as an art student, Landy had made textile works from junk materials. A revelatory encounter with the kinetic sculptures of Swiss artist Jean Tinguely – works that seemed to have 'no purpose' other than self-destructing – took his interest further.[80] Drawing on such influences, as well as his own experience of living through the economic upheavals of the Thatcher revolution, much of Landy's work between the 1990s and the 2010s was either taken up with the notion of throwaway culture – with waste destined for the landfill, an essential element of consumer society – or found ways to deploy junk and leftovers to devastating effect, to reveal how entangled our lives had become in a world of possessions.

The fact that such possessions became the principal means through which consumer society conferred a sense of identity on individuals was the inspiration to go to the lengths that *Break Down* would take him. This made it possibly the only work of its era in which an artist metaphorically and literally threw themselves into the work in such a strangely destructive and forensically detailed fashion, but precisely as an action in which creativity was expressed, paradoxically, through an act of dismantling, a kind of disappearing act. All of which is to say that the work, in this sense, was also a psychological exercise in unmaking, or detaching the self from the 'self-defining' objects that not only took the form of possessions, as detailed in inventories as well as sketched on a giant map of his extended self, but, as it turned out, included things that also in some way had possessed him. Landy would eventually become aware of the real self-annihilating consequences of making this type of work, insofar as he was turning into an artist who was not leaving much behind in the way of work that could be sold or, indeed, be possessed by the institutions of the art world or the various collectors who also conferred value within that realm.

But Landy felt that his works, as if acknowledging the efficacy of the things we become attached to, always seemed to 'assert their own destructibility'.[81] A publicity leaflet distributed by *Break Down*'s two co-sponsors, *The Times* and Artangel, carried the rationale for what some observers might have taken as another of the many contemporary art stunts that seemed to define the era of 'sensation' that was associated with the YBAs of Landy's generation.[82] 'From time to time we all enjoy a minor clear-out of our belongings,' the leaflet began, continuing:

> from those unwanted Slade Christmas singles, to clothes that are past their best. Michael Landy, however, is taking the clear-out to the extreme in his exhibition *Break Down*. Over the 14 days of *Break Down* Michael Landy examines our romance with consumerism by systematically breaking down everything he possesses. Every single thing . . . We are what we own. Over the past year, Landy (who lives and works in London) has made an inventory of everything he owns, every household good and every personal effect. He owns 7,006 different things.[83]

In the end, there may have been some confusion about the precise number of items. Some newspapers reported that there were 7,010 objects to be destroyed. 'With reference to your review of *Break Down*,' wrote one correspondent to the *Financial Times* after the event, 'Mr Landy will in fact be destroying only 7,009 objects in his exhibition . . . I was moved to steal one of them. It is my intention . . . to make my own little artwork out of this stolen object.'[84] Later reports stated that some 7,227 items were inventoried. The identity of the apparently stolen object was never revealed.

The 1990s was a decade of recession, during which 'once-respectable high street stores now provided temporary outlets for wide-boys selling dodgy stock in auction-like scams that enraptured small crowds of bargain hunters.'[85] Oxford Street, despite its location, was unable to escape unscathed. *Break Down*'s use of an empty C&A department store that had closed its doors to shoppers took aim at one of the main targets of his work: consumer society and its various spaces and emporia. Landy had earlier staged the recession-inspired exhibit *Closing Down Sale* at the Karsten Schubert Gallery in 1992. Another notable intervention was a 2010 commission to produce a work of art for a new London store opened by the luxury fashion house Louis Vuitton: in this case he produced a Tinguely-esque machine mounted on a wood chopper that was given the name 'Credit Card Destroying Machine'. The contraption accepted cards from anyone willing to give it a try.[86] In return the shaking and whirring machine, which was made of junk and scrap, was set into motion and produced a felt-tip drawing on paper already signed by the artist, which an assistant was on hand to give to the lucky shopper who had just sacrificed their credit card.

As people walked past the window looking into the space where the *Break Down* work was in process, they were often already themselves weighed down with newly acquired possessions from the many retail outlets on Oxford Street. Here, they could look in through the large plate-glass shopfront to see something rather odd: not the usual displays of mannequins bolted to the floor offering glimpses of the newest styles – 'adorned in dreams', to use the expression of fashion historian Elizabeth Wilson.[87] Rather, they saw Landy and his industrious helpers dressed in overalls and looking like operatives at a recycling plant, as they worked around a system of conveyor belts that carried everything he owned to a point where it would first be disassembled, then shredded and granulated in the manner of a waste reclamation facility – a place where things are typically broken down into component materials (metal screws, wire, plastics, metals, glass) so that they might be taken out of the waste stream. From the position of those staring in, it must have seemed as if the whole process that consumer society induced in individuals had been set into reverse on a strange stage set or inside a giant diorama come to life. People found it 'stirring', admitted a sceptical Matthew Collings, artist and art critic, and there was no shortage of stories of 'spontaneous cheering and tears' as Landy worked his way through the long ordeal of self-destruction.[88]

This was also the era, of course, when pop psychologists began to enjoy a burgeoning trade in books and articles on the subject of 'decluttering', something obviously related to the rise of self-storage spaces, as previously noted. On the surface, it might have seemed that what Landy was engaged in was a public version of those smaller and more domestic psychodramas: tidying up his life and clearing away the obstacles to happiness that the old material possessions represented. But it was much more than that. The existential dimensions were obvious to many who witnessed it, who would muse to the artist on the value of possessions of their own when it came to imagining themselves taking his place. 'People of the cloth', Landy later recalled, 'came to the exhibition and gave sermons.'[89] While he was able to claim – like other artists probably dating back to Marcel Duchamp, who also used the objects of everyday life as a means of escaping conventional aesthetic rules – that he wanted to disappear through his work, the aftermath of *Break Down*, when he was left with little more than the clothes that he was wearing, led Landy to a profound reassessment of his life and work. In the first place, there was the seriousness of the psychological impact of what he had done. 'It was like witnessing my own funeral,' he later claimed.[90] Second was the realization that throwing everything away amounted, if not to actual self-annihilation, then to a kind of self-sabotage. Having no work to sell as a result of his creative inclinations and process was becoming a costly habit. A later work titled *Semi-Detached* (2004), based around a replica of his parents' Essex home, garnered much interest when it was installed in Tate Britain, but it too had to be destroyed because it had been designed to be a one-off that could not be dismantled and reassembled. Nevertheless, it was not really Landy's intention to turn into 'The Man Who Destroyed Everything' over and over again. 'A lot of my work no longer exists,' he said in 2010. 'In a sense it is liberating, but as you become older, if there's nothing to look at, who's going to remember you?'[91]

Landy's work reflected a choice we may all have to face under conditions where consumer objects and personal possessions help to define who we are: namely, what to keep and what to throw

Michael Landy, *Break Down*, 2001, installation view, C&A building, Oxford Street, London.

away. The remains of *Break Down* that could not be kept out of the waste stream ended up in landfill.

THE IDEA OF THE landfill site as a final destination for the dreams that animated twentieth-century life is a key image in Don DeLillo's 1997 novel *Underworld*. DeLillo's often prescient writings saw him identified as someone who was able to discern hidden connections that revealed unacknowledged truths about America in that century. In the case of waste, we learn in *Underworld*, it was an idea that represented nothing less than 'the secret history, the underhistory' of that nation, an all-devouring phenomenon that seemed as if it could consume everything in human life. The professional apparatus of waste management is characterized in *Underworld* as 'an esoteric order' of 'adepts and seers' who are – unbeknown to the public – 'crafting the future': 'the city planners, the waste managers, the compost technicians, the landscapers, who would build hanging gardens here', or 'make a park one day out of every kind of used and lost and eroded object of desire'.[92]

The novel's principal character, Nick Shay, earns his living as a 'waste broker', an employee of a company – the suitably named 'Whiz Co' – that makes waste disappear. Reflecting on the quasi-transcendent and almost religious dimension that waste opens up, he describes the people in his line of work as 'the Church Fathers of waste in all its transmutations'.[93] Shay is also overcome by visions of the construction of sublime landfill sites that could swallow up the waste he sees all around, making it disappear forever: sites that would dwarf the pyramids as monumental icons of a culture. This thing, the landfill, would contain the evidence of what Americans had become by the end of the twentieth century, Shay believed. They would contain the stuff that revealed the deepest truths about the culture of the times.[94] Another character – the waste guru Jesse Detwiler – who imagined that there would one day be a new appreciation of the sacred nature of waste, thought that landfills could become sites that future populations regarded as shrines to the culture. 'Basic household waste ought to be placed in the cities that produce it,' he declared, and:

> Bring garbage into the open. Let people see it and respect it. Don't hide your waste facilities. Make an architecture of waste. Design gorgeous buildings to recycle waste and invite people to collect their own garbage and bring it with them to the press rams and conveyors.[95]

The landfill sites where we send the kinds of materials that we divest ourselves of are not like this, of course. Like the global waste that disappears to somewhere else, they are dispersed and out of sight to most of us. They are sites that constitute one of those non-human human zones, where natural processes may not only claim what was once our own but set in motion forms of environmental degradation that have sometimes transformed many older landfill sites into 'toxic-waste cleanup sites' – because of the presence of 'hazardous materials that should never have been disposed of in a landfill in the first place'.[96] Another danger relates to sites that may have been located in wetlands or other low-lying areas that put groundwater at risk of pollution.[97] But above ground, and before they are finally covered over and closed, the wastes that are moved around the surface also form part of another ecology, as a source of food for the birds that are often seen in images of landfill sites.[98] The population of gulls, Tim Dee notes in his fascinating book *Landfill* (2018), 'boomed through the throwaway decades of the 1960s and 1970s' before the 'recycling or incineration of food waste' began to make 'edible trash' scarce.[99]

But it is the very fact that these places are far enough beyond the spaces we inhabit – or obscured from the world as we move through it – that allowed them to proliferate. The 'layers of soil' that are used to cover up the stuff that is pushed into these holes could be seen – much like the underground world that enables life to continue as normal above ground – as a metaphor for our social amnesia when it comes to the material wastes of everyday life.[100] In modern times, wastefulness – which is usually perceived as a moral deficiency of some kind – has been driven by the fact that what is out of sight is also out of mind. This could even be described as a logic of waste creation, which is the other side of the demand that consumer capitalism, in making its abundance so visible, stokes for its goods and products; a demand that is likewise hidden in the secret (and maybe not-so-secret) desires of the consumer, sparked by the 'synergistic action of major cultural objectives such as comfort, aesthetic quality, individual choice, and novelty'.[101]

The Litter Craft

Picture the scene. Fallen couches and broken chairs rest by roadsides and pavements, some looking to all intents and purposes as if they might have moved there of their own accord or still be in transit, merely taking a break as they inch towards some unknown destination. Looking around, one sees patches of ground with surfaces of rough, untended grass. Spaces and corners are strewn with tin and plastic leftovers that one can easily imagine coming to life and creeping back towards pathways where people otherwise pass by, oblivious to the stuff. Squashed cans and plastic bottles hug gutters at the side of roads, while lost supermarket trolleys come to the surface on muddy canal banks and lonesome refrigerators, scuffed and bashed, stand to attention (where they have not fallen down), asserting the fact that they are no longer wanted or desired.

Into this zone of the 'in-between', Dauvit Alexander and Dan Russell, guerilla jewellers based in Britain's second city, Birmingham, went in search of stuff that seems to find its way to such interstitial spaces when no one is looking, intent as they were on bringing some form of redemption to the inert, apparently dead things that they had begun to notice materializing, scattered on the ground all around them. In seeking out improbable materials for their new jewellery collection, they were drawn to the type of detritus that is ready to spoil a favoured view, or perhaps clog up the passageways between our destinations as we negotiate the routes of our daily travels through such seemingly abandoned spaces.

Such patches of urban wasteland are, of course, imaginatively constructed zones that we mostly perceive to be in some way unpoliced, unregulated or – reflecting an older meaning of 'waste' – constituting a thing or a place that belongs to no one. One might illustrate an example of the indeterminacy of ownership and property regarding that which has been thrown away through the phenomenon of 'freeganism', which gives people who don't like the idea of food going to waste a licence to appropriate goods thrown away by supermarkets (once such goods are put in places outside the space of the stores they formerly belonged to). Perhaps the sense that what is placed outside is free to use – no responsibility claimed, no possession taken – explains why such scrubby patches of land exert a near magnetic pull as locations where all manner of unwanted things are dumped. Urban wastelands thus become the scene of multiplying waste; a dumping ground for other unwanted stuff.[102] In patches of land that can be found around Tokyo, writes anthropologist Peter Wynn Kirby, one sees unusual acts of dumping that reflect this fact in very strange ways, including the 'abandonment of pets such as cats and dogs', which in some quasi-magical act of exchange leads to 'an abrupt rustication of these domesticated pets in a simultaneous dilution of "the wild"'.[103] Such examples illustrate in one way how waste finds the spots where it seems to proliferate, how it wends its way into spaces and places we associate with the mobile and indeterminate, reflecting in turn how 'waste' then becomes a category that could absorb almost anything that is in transit between accepted categories of value or understanding.[104]

And, in a way, that thought helps to illustrate how the *stuff* of waste itself embodies a mobility and indeterminacy: that it is one thing that becomes (or has become) another. On the one hand, waste is in flight from accepted categories of value, and on the other it moves towards – at least to those who have dispossessed themselves of unwanted things – a condition of total worthlessness. In this sense, waste, occupying a conceptual space between existence and non-existence, has distinctly metaphysical overtones: it directs our thoughts to life and what lies beyond it (even if we are 'merely' contemplating the afterlife of rubbish).

Throughout history, by contrast, and as our jewellery scavengers recognize implicitly, not much has been as consistently valued or cherished as the things with which people adorn their bodies. This is especially true for those very personal objects – bracelets, necklaces and rings – that we become intimately attached to. But such items of jewellery – unlike much of the other stuff we might desire to possess – aim for a permanence

Three 'Flower Bomb' pendants, made by Dauvit Alexander, 2023, from nitrous oxide cylinders, silver, synthetic and natural gemstones and cold enamels.

of attachment that moves them into a category of the precious and valuable, whatever the personal basis of that value might be: the broken keepsake inherited from a deceased family member can hold incalculable value because it is, as Sherry Turkle would describe it, an 'evocative object'.[105] Such objects would seem to belong, in fact, to a consciousness that places emphasis not on speed or instant gratification, but something close to the opposite: an appreciation for things that are intended to be permanent rather than disposable. For these kinds of reasons such adornments also – and more conventionally – serve as tokens of commitment, and as such, those who design and make jewellery could be expected to understand that the difference between the 'throwaway society' and their own practice is in making things that have some lasting value. It is true that, in addition to its symbolic worth, jewellery may be valued because of the materials it is made from. On their own, substances such as gold and silver possess intrinsic economic exchange value. But what gives the jewellery of Dauvit Alexander and Dan Russell the possibility of enduring is that the objects they craft become examples of the one-off, the unique and unrepeatable, made from an assortment of unwanted and discarded materials that might not be easy to acquire again in the exact proportions that are required to go into the making of, say, a bracelet or necklace.

If we can learn anything from the patches of urban dereliction and abandonment that offered up the materials that were used in this jewellery, the places that seem to have some magnetic attraction to litter, it is that they provide an illustration of how fast life moves and how this acceleration reveals itself in the form of stuff that is literally cast off by the harried or unthinking pedestrian. Yet another contingent by-product of life, this litter comes to provide essential insights into who we are. Arguably, we long since arrived at a place where the necessary accompaniment to the acceleration of life is experienced in part as more *stuff*. In that respect, production and consumption are caught in some infinite repetitive loop – but one that can nonetheless be disrupted by taking up residence in the mutable space of what we might term the 'in-between'.

3

Resources: Reclaim, Recover, Recycle

In modern societies, time is understood as a progressive phenomenon. It accords with a view of history that at its most idealized sees us overcome all defects of the past. Waste, in all its variety of forms, whether taken individually – in terms, so to speak, of first appearance – or in combination, as a spectral totality of the kind that may disturb our thoughts when we glimpse some unknown or suppressed dimension of it, contradicts that progressive idea because it locates us within other temporalities that seem to exercise a certain claim on us.

A Time to Waste

There can be no history of waste that charts the passing of time as a path of victory, illustrating how we conquered one of its forms before tackling and defeating the next, and so on. Even to speak of 'forms' of waste isn't quite right, for much waste is either mutating into formlessness or is in transition to another state, which is surely the case when certain waste materials are processed or recycled and take on new and maybe different forms that make the materials useful once again. The sphere of waste nonetheless only contains things that multiply and become more variegated over time and history. Without constant vigilance and the investment of substantial time and energy to keep it at bay, waste in all its different forms can easily overwhelm us. We might wonder, then, what role time actually plays in our perception of waste.

From the ubiquity of clocks and timekeeping devices to the idea that we are living in or through a period of climate emergency, time – specifically as a phenomenon that connects us to uncertain futures – seems to make claims on us, if not to press its will upon us. Our sense of the unknowability of the long future in particular makes time, both in abstract and concrete terms, something that is always present; 'a powerful force in our imaginations and daily lives'.[1] To understand the anxieties that result from the complex ways that our lives are tied to certain temporal logics, we need to see how these are derived from much earlier, premodern ideas about time. These relate specifically to what has been termed the 'Christian invention of time', which occurred in the period of late antiquity (*c.* AD 300–500), when time began to be conceptualized in terms of notions such as eternity, life after death and the end of days.[2] Just as we live today with the prospect of ecological destruction weighing on our minds – an idea of ends that reshapes how people understand present concerns – so too 'anticipating the end of days' was a 'deadly serious concern of early Christianity, a concern which changed how it was thought possible to engage in everyday life and its continuity'.[3]

It is from such origins and through social changes that occurred among religious communities of northwest Europe in the Middle Ages (and specifically through the cultures of a variety of Benedictine monks, Protestant reformers and Puritans) that we have inherited ideas of time-saving, time-keeping and time-rationing as means of avoiding the squandering of what we now more broadly regard as a precious resource, namely time. The waste of time, in other words, is a malady that became fully manifest – in a way that we in the present can relate to – in the Middle Ages. And it was in the medieval monastery, 'the great focal point of civilization in the early Middle Ages', that a

Time can be an elusive resource, and whether spent or not may still be wasted.

practical and economic revolution in life – according to German sociologist Max Weber, writing at the beginning of the twentieth century – planted the seeds of the modern world. Thus we find that modernity goes hand in hand with a cultural attitude that regarded wasting time as the greatest of sins.[4] The monasteries of the period in which this attitude was incubated had a profound influence on their societies and were much more than just focal points for spiritual life: they were communities that facilitated the preservation and transmission of culture (through the creation and preservation of texts and the establishment of libraries), managed land and estates through the use of the most advanced technologies of the time, and became, through the labour of monks and other dependents, hubs of production as well as focal points for spiritual life.[5]

Their existence as isolated communities in the countryside, however, reflected something of the precarious nature of the civilization of the Western world at that point in time, following the fall of the Roman Empire in the fifth century.[6] 'Many towns slowly but inexorably faded away,' writes Vito Fumagalli, 'and economic and social life died as communities became isolated from one another.'[7] In short, the decline of the 'networks of communications and relations of the ancient world' that had spread over the lands of Europe with the advance of Rome left a civilization of 'isolated points, of oases of culture in the middle of "deserts", of forests and of fields returned to "waste", or of countryside barely brushed by monastic culture'.[8]

The monastery, in Weber's analysis, was 'the model of the rationally administered agrarian and commercial enterprise'.[9] And it was from these dispersed centres of culture that came ideas that would go hand in hand with the emergence of new means of marking and saving time.[10] The experience of time, for the most part, was as something external and environmental, an agricultural time that inhered in nature and its seasonal changes and that entailed a lot of waiting, but was certainly something that few could measure or manage.[11] The masses, Jacques Le Goff notes, 'obeyed the time imposed on them by bells, trumpets, and horns', which sounded out and filled the atmosphere as calls to attention and action.[12]

By contrast, there would be no modern society without a conception of time as something subject to human intervention in more and more detailed ways. Modernity itself, of course, was a consciousness of time as something that structured life in certain ways, as opposed to the understanding of time as something that was found in nature and the seasonal rhythms of natural life. Lynn White Jr has noted that illustrated calendars from around the early ninth century, well before the invention of mechanical clocks, were already changing to reflect a different human relationship to the land as a resource, and showed the year ahead as a time filled with work to be done:

> In older calendars the months were shown as passive personifications. The new Frankish calendars, which set the style for

> the Middle Ages, are very different: they show men coercing the world around them – plowing, harvesting, chopping trees, butchering pigs. Man and nature are two things, and man is master.[13]

In a world without such mechanical means of marking and calculating time, to think of the future was to think ahead in a much more circumscribed way than we are used to now. Today we think of time and history – the passage of time, time past, and historical events – in terms of notions such as progress and decline. We live in cultures in which growing older is often accompanied by nostalgia for some golden age that has just passed in our own lifetimes.[14] 'Before the invention of mechanical clocks in the thirteenth century the question, What time is it? was not very urgent,' writes Svetlana Boym. 'Certainly there were plenty of calamities, but the shortage of time wasn't one of them.'[15] No shortage, no concerns about wasted time. An illustration of the new importance of time in the Middle Ages appeared in the early fourteenth century as an allegory in a widely distributed text titled *Horologium sapientiae*, or 'Clock of Wisdom', written by a German Dominican friar, Heinrich Seuse (Henry Suso). It shows the figure of Wisdom pointing towards a mechanical clock and instructing a monk, who is seen taking notes on the importance of time and, we might infer, the necessity of not wasting it.

The invention of the mechanical clock and its effect on life marked a change in consciousness. 'From the first half of the fourteenth century on', Le Goff observed, the idea of time running away 'became more specific and dramatic', and 'wasting one's time became a serious sin, a spiritual scandal.'[16] Indeed, to be oblivious to time's passing, and to *waste* time, was to be considered less than human; a view summed up in the teaching of a Pisan preacher named Domenica Calva who, as Le Goff notes, had written that 'the idler who wastes his time and does not measure it was like an animal and not worthy of being considered a man.'[17] Calva's contemporaries could be found secluded away from the towns and cities of medieval Europe, where they developed new cultural attitudes and social practices that fundamentally gave a modern moral dimension to the understanding of 'waste' as an idea that had its roots in a certain kind of spiritual life. As Max Weber wrote in *The Protestant Ethic and the Spirit of Capitalism* (1905), wasting time was, for the ascetics who viewed their time on earth as God's time, 'the first and in principle the deadliest of sins'.[18] Weber's seminal work explored the nature of what made the medieval monastery and its monks – methodical conservers of time and resources – exemplary of the rational and calculating approach to life that would give birth to a modern economy and the attempt to conquer nature. While it was still true that the acquisition of wealth could lead the pious to fall from grace, the Calvinist Protestant view that the individual was 'only an administrator of what God' had bequeathed gave rise to an idea of the 'calling': to serve God through 'activity carried on according to the rational capitalistic principle, as the fulfillment of a God-given task'.[19]

This change in time consciousness prefigured the transition from an agrarian to an industrial society. It was an attitude carried beyond the orders of monks into 'the professional ethos of Puritanism', which in turn expressed or gave form to a 'secular version of the ascetic ideals of the monastic life'.[20] But we may even identify it as a general condition of consciousness, certainly in Western consciousness, because upon closer consideration, our contemporary preoccupation with waste usually reveals itself to have a temporal dimension: it revolves around questions of ends, outcomes or consequences and their relationship to how we live in the present. Such an outlook reflects an attitude or ideal of continual improvement and progress; an awareness of how easily resources or untapped potential may be squandered, if not also of the unintended consequences of failing to remain diligent and in control of our lives. Medieval Christianity's idea of irreversible time – a belief that we were caught in the decay of a temporality that was in eternal flight – became deeply ingrained in Western thought in the centuries ahead and informed all manner of future-orientated secular and utopian projects. And while utopian thinking can be seen as a manifestation of the tendency to view the past as itself dispensable – something that is often only fit for the metaphorical junk heap – such thinking would hardly be 'conceivable outside the specific time consciousness of the West, as it was shaped by Christianity and subsequently by reason's appropriation of the concept of irreversible time'.[21]

The spectre of wasted time, in its identification of a resource to be harnessed for some present purpose or greater good, came to be reflected in attitudes to the land. The eleventh-century Domesday Book could only measure the extent of what we would call wasteland (that is, land unoccupied or untilled, sometimes described then as 'silva') in terms of its capacity for supporting pigs; which is to say, in terms of productive potential. By 1696, an English statistician named Gregory King – in his *Natural and Political Observations upon the State and Condition of England* – was able to estimate that the wastelands of England and Wales accounted for more than a quarter of the land in the country. The reclamation of wasteland in England, according to environmental historian W. G. Hoskins, was under way by then, having begun in earnest during the thirteenth century.[22]

At around the time King made his estimates in the late seventeenth century, there appeared to be a general concern within Christian thinking that the land and the fruits of nature were not simply given as a consequence of some existing state of abundance. In fact the belief was quite the opposite: there was no such abundance – instead there was only the intimation of land gone to waste – and, as one commentary on the biblical creation story puts it, without the vigilance of God, expressed through his stewards on earth, 'the order of nature would be wiped out in a moment and would revert to the original chaos.'[23] One might compare this belief to the earlier sentiments expressed in Thomas More's *Utopia* of 1516, where the inhabitants of the fictional island of the same name are depicted as ruthless advocates of the practice of claiming land that was going to waste, thinking it 'justifiable to make war on people who leave their land idle'.[24] As well as its impact on those parts of the British Isles that seemed to be unproductive, the belief that land that was not worked could belong to anyone who laid claim to it or made it productive in some way led to colonizing efforts across the lands of the British and other European empires.

By 1800 efforts to reclaim wastelands – through the use of methods such as pumping and draining, and also the enclosure of common land – reduced what was also simply termed 'the waste' to approximately one-fifth of the total area of England and Wales.[25] Water had to be channelled and put to use or else, as Joseph Amato writes, 'the earth would remain an untamed surface of swamps, damp ground and foul wastelands.'[26] The rationally calculating and time-conscious sense that waste was in truth an untapped resource was in tune with the philosophical wisdom of the age, as expressed by Gottfried Wilhelm Leibniz in his *Monadology* (1714): 'There is nothing waste, nothing sterile, nothing dead in the universe.'[27]

WHILE IT WAS CLEAR to thinkers like Leibniz that the physical world contained potential for improvement, it was much more complicated to say the same about time, which he saw as just an 'order of successions' that could only be given form when seen in connection to the materiality of worldly existence. When it came to conquering time, monastic culture showed one of the ways in which waste would become a strange presence in the modern world. For the contemporary German philosopher Peter Sloterdijk the medieval monastery was a 'container for boredom', and it is perhaps in the idea of boredom that we can make a connection between an attempt to control a temporality that appears as a runaway agent of disorder and the various temporal disorders and anxieties of the present.[28] Fast-forwarding to contemporary life we can see that these cultural influences, developed over centuries and inculcated in one generation after another, had been diffused or disseminated through what the sociologist Eviatar Zerubavel terms an

> 'activity cult', whereby people are expected to maximize their 'active' time and to minimize any 'empty, unaccounted-for' time periods . . . The popular use of colloquialisms such as 'spending,' 'wasting,' 'saving,' 'investing,' 'allocating,' and 'budgeting' with regard to time is quite indicative of this trend.[29]

In the industries of the twentieth century this 'cult' notoriously gave rise to time and motion studies, something that I myself was once subjected to as a young apprentice gardener when, alongside an older and more experienced colleague, a man with clipboard and stopwatch looked on and took notes as we trimmed a long hedge that must have been 45 metres (148 ft) from one end to the other and over 3 metres (10 ft) tall: me with a foot on the bottom rung of a ladder to hold it steady and my workmate going up and down its steps, shaving this behemoth with a power trimmer. It was through such methods that the rationalization of factory work had already been taken to a peak of efficiency early in the twentieth century in the U.S. factories of Henry Ford, where the Model T vehicle production line was designed to waste no effort, even going so far as to create tasks that could be fulfilled by those who might have been otherwise deemed to be trapped in unproductive bodies. As Ford noted, the production of his vehicle required '7,882 distinct work operations'.[30] Lest there be a shortage in the able-bodied who might have had first call on such work, he declared that there was no space for waste on the production line, and it had therefore been determined that of the total number of operations required, '670 could be filled by legless men, 2,637 by one-legged men, two by arm-less men, 715 by one-armed men and ten by blind men.'[31]

Garbage Reconfigured

Sitting within a few feet of a figure that I take to be a real-life manifestation of Nick Shay, the waste management consultant of Don DeLillo's *Underworld*, I find myself in the darkness of a room that welcomes visitors to the twilight world of waste, to the space of what I would term the 'in-between'. This is conceptual space wherein waste is considered, speculated upon and reimagined as something – be it a thing, a property, an idea, a potential unrealized – that represents merely a temporary problem: one that, like the materials used in the making of 'trash' jewellery, fluctuates between the always subjective conditions of value and worthlessness. But this is a space in which what is required is also a measure of certainty. In the world of waste, time is at a premium. This is why, in *Underworld*, Shay ends up compelled to travel the world to speak at conferences about 'the intractability of waste': its potential to drive human society one way or the other – towards possible collapse or towards greater ingenuity of means in combating the problem of waste – and the spectral presence it manifests in the world of human goods that he believes remains obscure to most people. It is in just such a setting, a 'Waste Expo', that I learn about waste exchanges; about the startling claim that it is possible, to quote my Shay stand-in, to 'take the phosphoric acid used in testing microprocessors and put it in a can of Coke' and thereby 'improve the product'.

Such events are now commonplace as occasions where the waste management industry, its numerous company and organizational delegates and potential customers, gather. This is the kind of place where the latest innovations in a fast-changing sector of the economy are discussed, products and services are marketed, and various myths about business wastefulness are slayed. The message seems to be clear: there is no waste, only potential resources as yet unrecognized for their value. Writing ahead of his time in 1982, the English architect and critic Martin Pawley would similarly observe that it was premature to condemn material things that had fallen out of use to the status of waste. 'Matter is indestructible,' he declared, suggesting that 'the creation of waste is in reality the incomplete transformation of matter.'[32] In a 1975 book titled *Garbage Housing*, Pawley looked at what could be learned about continuing the transformation of matter in 'secondary use', looking at the variety of dwellings that people had improvised from waste materials. Some of those same materials were of the

Glass bottles repurposed into building materials, an example of 'secondary use' advocated in Martin Pawley's *Garbage Housing* (1975), Cap-Egmont, Prince Edward Island, Canada, 2020.

sort found in shanty-town dwellings and improvised shacks that existed and provided shelter the world over for those who could not afford something better. Pawley wanted to carry forward the lessons of this kind of building into thinking about a more systematic response to the disposability of certain materials and how they could be developed in future by designing *into* them (which amounted to changing the way we thought about them) an idea that they could be used for more than one purpose. This was the idea of secondary use.

Pawley suggested that 'garbage housing', which consisted of materials like corrugated metal, product containers, bottles, oil drums and indeed anything else that was suitable, was more common than we generally knew and could be found on every continent of the world.[33] For Pawley, the case for promoting the idea of houses made from newly designed recyclable materials – such as the Heineken beer bottle that when emptied of its contents could be reused as a glass brick – was one that had to be made within the context of a need for affordable housing. This was at a time, in the UK, of property market inflation and during a period when the practice of squatting (the taking over of abandoned or condemned properties, even entire streets) was becoming more common, and homelessness was increasing. Dwellings, Pawley argued, had lost their real meaning and significance and instead had come to be seen as something akin to containers for the products of consumer society; the home was, 'in effect, a consumer envelope'.[34] The challenge was to rethink the purpose of the house, to reconceptualize it, and at the same time try to devise ways of making the practice of housebuilding less expensive by embracing the idea of secondary use.

In 1977 Pawley constructed an example of one of these garbage houses with students he was teaching at the Rensselaer Polytechnic Institute in Troy, New York, where his ideas had led to him being offered a position as a visiting professor of architecture. The house they built utilized a variety of unusual, discarded materials: six hundred thick cardboard tubes left over from newsprint press rolls were used to establish the frame of the dwelling, for example. Also included were miscellaneous bits of scrap material and a supply of non-returnable 16-ounce bottles, which Pawley assured the *Los Angeles Times* were sturdy building materials, given that they had been designed (as containers to carry liquids) 'to withstand greater stresses during pasteurization and capping than bricks or concrete blocks undergo during normal construction loading'.[35] The walls of the house comprised 2,000 large No. 10 cans (capacity 3 quarts, or roughly 3 litres), on top of which sat a roof made of layers of corrugated cardboard sealed with coal tar and waste rubber and insulated with scrap cotton polyester. The whole point of all this, as Pawley explained, was to show that, instead of taking out loans to buy homes, people could actually build their own basic house using the methods and sorts of materials he had used for a cost of around $1,000 (at a time when the average new single-family home in America was about $50,000).[36]

This idea, though, seemed destined to meet a host of practical problems. Where conventional home construction was supported by the use of mass-produced standardized components delivered through long-established supply chains, how easy would it be for the average person, or even a group of people or a family working together, to find or collect the materials required for the construction of a garbage house? How were problems with integrating water and electricity and other utilities to be solved? How would the attitudes of neighbours reacting against a garbage house in their midst be overcome? In the end, the *Los Angeles Times* suggested, it seemed too soon for garbage housing. 'Problems of collection and distribution complicate American life. Solid waste will only serve the masses when there are massive storage and delivery systems to facilitate its reuse.'[37]

One might be tempted to describe Pawley as a 'recycler', and in a certain sense, it might not be a bad description. He was extending the life of materials that had been seen as having little value. In doing so, his example helped to challenge prevailing attitudes to material waste. But at the same time, as the creator of new forms and structures, even when they exemplified what might be described as a cheapskate aesthetic, Pawley was not recycling in the way that we understand the concept today. Recycling, in that latter sense, refers to something much more specific that arrived as a new means of restitution at a certain point in the history of the twentieth century, involving a more conscious effort that was very broadly directed outward and at restoring some balance between the human use of resources and the effect that this could have on ameliorating the environmental consequences of resource use in societies driven by consumerism.

Recycle

Writing in 1971, the geographer Yi-Fu Tuan attempted to outline the sources for a developing environmental consciousness.[38] In the wake of Rachel Carson's *Silent Spring* (1962), published almost a decade earlier – which drew attention to the reckless use of pesticides that were spoiling natural environments and being linked to health problems in society – Tuan looked at a number of recent book-length historical studies of the relationship between humans and the natural environment.[39] Where Carson's work had developed out of her observations as a marine biologist, Tuan raised the question of the importance of cultural and historical work in understanding how 'environmental attitudes' had shaped the human–nature relationship at other times. It was the kind of work, new in its subject and approach, that indeed signalled a shift in sensibility and one that might cause 'a future historian of ideas to look upon the late Sixties and early Seventies of the twentieth century' as the era when concepts like 'environment' and 'ecology' entered mainstream discourses and were brought to prominence to highlight issues that would come to shape the political climate of the future.[40]

A few years before Tuan's observations, the American historian Lynn White Jr had noted that language played its own part in how we thought about our relationship to the natural world, and that one way of seeing this was to recognize how recently it was that 'ecology' – a word, he noted, dating from as recently as 1873 – had become a concern in human affairs. 'As we enter the last third of the 20th century,' White wrote in an article titled 'The Historical Roots of Our Ecologic Crisis' (1967), 'concern for the problem of ecological backlash is mounting feverishly.'[41] By using this language, he was suggesting that nature was eventually going to hit back. In 1970 Earth Day was founded, and it would be the first time that attempts had been made to mobilize people en masse around environmental concerns.

White's 'Historical Roots' article had also located the origins of late twentieth-century environmental problems within a much broader context, specifically the medieval beginnings of technological society, which we could identify with the rise of industrial society and the subsequent growth and spread of capitalism. In particular, he suggested, the Christian culture of Western societies inculcated an idea of progressive time, which was the source of technological advances that submitted nature to forms of human exploitation, just as Calvinist doctrine commanded. Christianity thus 'not only established a dualism of man and nature but also insisted that it is God's will that man exploit nature for his proper ends'.[42] Among those who were agitating for some response to the developing awareness of pollution and other ecological consequences of population growth and resource exploitation, there was another new concept emerging at the turn of the 1970s: recycling.

At that time anti-waste drives would have been part of the shared memory of those who had already lived through the world wars. These, of course, were driven by the urgent needs of the times, the sense of emergency or existential threat. In 1916, for instance, a 'National Economy Exhibition' toured around Britain, revealing the many and varied ways that food and other material resources could have their lifespan extended. To save waste was to 'economize'. One display, preserved in pictures of the event, looked akin to a contemporary crime scene exhibit. At its centre was a domestic dustbin lying on its side, its contents spilled over a table and tagged with labels containing information about how the various items might be rescued for other uses. 'Saving' indeed – not recycling – was the watchword then, as one report at the time revealed:

> Anyone who understands the rudiments of economy . . . knows that what saves labour and time also saves money. Money that can be saved without discomfort is money saved

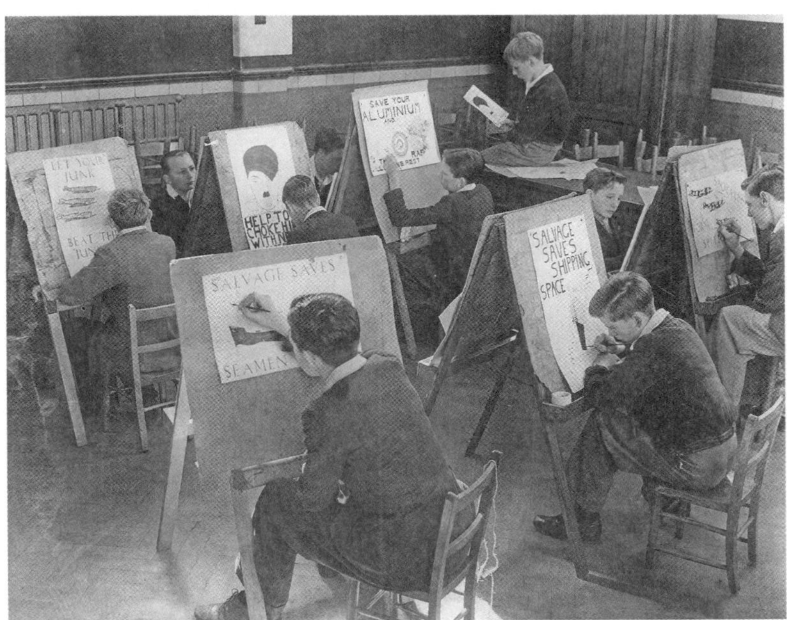

London school pupils painting posters for use on their own waste-collecting systems, c. 1939–45.

> from unproductivity. It is released for productive uses, and therefore for the good of the commonwealth.[43]

Information about resource use, deployed here for an educational or consciousness-raising purpose, was always a key component of these waste-minimization drives. Waste-themed advertising campaigns, unsurprisingly, proliferated during times of war; this was not only for the ends of salvaging materials or substances that might go into the making of ordnance or machinery, but to recover the waste that was produced within the domestic sphere, whose other potential military uses might not have been known. This style of campaigning created its own peculiar consciousness during the years of wartime scarcity.

During the Second World War, the U.S. Office of War Information was exceptionally active, employing graphic art to put together striking combinations of words and images to drive home its message: 'Save Waste Fats for Explosives', 'Food Is a Weapon, Don't Waste It', 'Eat Less, Waste Nothing', 'War Time: Don't Waste', 'Don't Waste Gasoline, Rubber, Money: Share Your Car', and others. What these illustrate is the extent to which the wartime economy of the United States, insofar as it related to the consumption of resources, was reconfigured around ideas of restraint, efficiency and vigilance, with overindulgence and carelessness represented as life-threatening and wasteful behaviour due to the broader impact they had on the war effort. For many people whose formative years were spent living in these conditions, the shackles of austerity were hard to throw off.

At the turn of the 1970s, advertising hoardings and billboards backing campaigns by the Environmental Action Coalition in the United States sought to demonstrate how to move from idea to action, and often carried echoes of those earlier wartime emergency campaigns. The messages they contained were intended to suggest that waste represented a looming environmental crisis

A young woman publicizing a wartime salvage campaign, Long Beach, California, October 1942.

while also grabbing the attention of people who might be otherwise absent-mindedly doing exactly the opposite of saving and reusing. The aim was to show that the materials that people had become accustomed to throwing away could, if properly managed, be transformed into some new resource. 'Don't Waste Waste' was the simple message of one statement, which underlined its injunction with the new idea: 'RECYCLE'. As well as prominently displaying this term, other adverts for recycling services featured instructions on how to prepare and sort the materials that would become the focus of the early recycling industries – glass bottles, cans and newspapers – in order that they could be dealt with efficiently.

Anyone who wondered where this new consciousness might lead could have looked at Ernest Callenbach's 1975 novel *Ecotopia* and believed that it was a work of science fiction describing a state of affairs far from the reality of the times. The story fast-forwarded the ideas that were beginning to sprout in the late 1960s and pushed them towards the end of the millennium (the story is set in 1999) and into a future where the idea of recycling was now a mundane facet of life in an ideal society that had seceded from the United States in order to address the impact that humans were making on their natural environment. Callenbach's book, in the words of Fredric Jameson, 'provided a summa of all the disparate sixties Utopia impulses', pitching itself as the source of ideas 'around which a whole political movement might crystallize'.[44]

In *Ecotopia*, Callenbach's narrator is a journalist on assignment from the USA, who travels to Ecotopia to gain better insight into the ways in which the ideas that first became widespread in the late 1960s and '70s had been logically drawn out to create a new society in which the maximum reduction of waste had been achieved by abandoning most of the goods and materials that modern societies had built themselves around. Biodegradability underpinned the 'stable state' concept on which Ecotopia, the place, was established.[45] The sense of the novel being ahead of its time is clear to see in the descriptions of goods and products designed with hi-tech materials that would simply decompose into the earth after use: for instance, plastics that were used in a variety of packaging types had 'a short planned lifetime and would automatically self-destruct . . . after a month or so, especially when exposed to sunlight's ultraviolet rays'.[46] Glass or pottery would simply disintegrate into sand. These kinds of ideas, as we will see, are echoed in the present day, in the notion of 'zero waste' and its aim of encouraging innovations in design to solve the problem of waste.

IF WE BEGIN TO understand the origins of post-1970s recycling and see it as a new concept – a way of looking at the human–nature relationship – that arose in response to consumer society, it raises the question of what we say about the use and reuse of resources in the pre-'recycling' world. Because, as we have seen, there was no shortage of means for recovering, saving or reusing unwanted and leftover materials or products before this new waste consciousness, and the processes it gave rise to, began to become apparent. We may be inclined to think that 'recycling' materials identifies a revolution in practice rather than in consciousness, and that what it represents – which we might also describe in terms of older ideas and concepts of saving and reusing – is something unique to contemporary society. Nothing could be further from the truth. Yet at the same time, the fact is that the term 'recycling', as used to refer to practices we are familiar with today that relate to a rationalized system of materials management, seems to be almost entirely absent in the English language before about 1970.

The *Oxford English Dictionary* shows that uses of the term 'recycling' before this point in time occurred in contexts that indicated something quite different to the associations it holds for us today. Today we understand it as a concept that enjoins a personal or civic responsibility for managing the kinds of waste that emerged in the era of mass production that we are personally responsible for making through our consumer habits and choices. By the 1970s, when this new understanding of recycling began to appear, it was not uncommon to still find the word used within the context of industrial recovery and so-called 'laundering' – or chemical recycling – which was a line of business that had been operating in the UK since the 1930s. There was money to be made, for instance, in the reclamation of oil as a by-product of industrial processes, but less so in respect of the unknown quantities of the stuff that was spilled or poured away in the millions of places where private individuals made use of it (in home garages, for instance, where a bucket of oil recovered from engines after a

This 1970 American poster reflects the arrival of the new concept of recycling.

change of lubricant could be of use to those who repaired their own vehicles).

In the pages of the British press, long before contemporary waste expos and conferences, the idea of a 'waste exchange', where merchants could meet in public and trade, much in the manner that the old corn exchanges permitted, was suggested in the 1860s.[47] But it was not until 1974 that the UK Department of Industry finally established such an exchange to deal in waste materials from manufacturing processes in general, which is to say all industrial wastes except for those that already had specialized processes or services associated with their disposal.[48]

Where did that other non-industrial waste go in the age before recycling bins and containers? It went to scrapyards, street traders and flea markets, to jumble sales, pawnbrokers and second-hand dealers in everything from TVs to antiques; to book, record and tape exchanges, to dealers in vintage or what is now termed 'pre-loved' clothing, charity shops, yard sales (in North America) and the roughly equivalent opportunity offered by car boot sales (in the UK).[49] Organic or food waste went to feed animals. As Mike Davis recounts in a fascinating section of his 1990 study on the rise of Los Angeles, *City of Quartz*, the success of Fontana Farms in that city in the first half of the twentieth century – a giant enterprise that produced food for millions – was, in large measure, the result of a successful application of a circular ideal. For decades, 'five or six hundred daily tons of garbage' – largely organic matter – from Los Angeles arrived by train and 'fattened the sixty thousand hogs' that made Fontana Farms the largest example of 'vertically integrated, scientifically managed, corporate agriculture' in the world:

> When the hogs reached full weight they were shipped back to Los Angeles for slaughter, recycled garbage thus providing perhaps a quarter of the region's ham and bacon. The co-incident accumulation of manure was no less valued: it was either utilized as fertilizer for Miller's citrus grove (also the world's largest) or peddled to neighboring ranchers.[50]

Across the Atlantic, the waste produced by the average British household up to the 1970s was, for the most part, contained in a single steel bin and often consisted mainly of ashes and cinders. In a time when coal power was still the predominant method of heating homes, much waste was simply burned in the home (cardboard, newspapers and so on), making the domestic fireplace a 'waste management technology' of sorts, and one that consumed anything that would burn, including 'rags and putrescible matter including waste fats'.[51] Other leftovers were often exchanged for reuse (especially in the case of bottles, which would return a cash deposit to the holder, something that eventually became economically unviable, though contemporary schemes of this kind have made a comeback in some places). Equally commonplace was recovery by 'rag-and-bone collectors, a dying breed who in the decades following the Second World War seemed increasingly to belong to a bygone world. But it was an occupation that had nonetheless persisted for centuries until the rise of the throwaway era, its decline hastened by other factors, including the gradual disappearance of the housewife (from whom they would buy unwanted items and who typically had, until the second half of the twentieth century, a stay-at-home role that was focused on household management).[52]

Collectors and Flea Markets

In the era before the birth of the throwaway society, as Susan Strasser notes, rag-and-bone collectors formed part of a broader class of street scavengers who constituted an informal system of waste management: 'Itinerant rag-and-bone men worked the streets of urban neighbourhoods, offering cash to poor housewives and selling their wagons full of refuse to dealers with storage facilities.'[53] Strasser's work relates mainly to the situation in the United States, but the practices she describes were common elsewhere. In Paris, the *Manchester Guardian* reported in 1924, rag-and-bone families – whose children grew up in the trade and gained a 'special flair for the value to be found among the wastage' – in one year had extracted from the vast quantities of rubbish heaps around the city between forty and fifty million francs' worth of materials to mend, reuse or sell on: cotton, linen, paper, string 'and vast quantities of bones, tin, glass, and metal'.[54] We might suppose that certain practices, such as the collection and reuse of bones in the making of tool handles and other functional implements, go back a very long time into prehistory and form part of the origins of human tool-making.[55] But the emergence of the rag-and-bone collector as a distinctive figure in the management of certain wastes can probably be more precisely located.

In the fourteenth century, it is thought, the collectors who had long travelled the English countryside, going from one town or village to the next gathering animal bones to be ground up for use as fertilizer, began taking the unwanted rags of a new material, linen, which was made from flax. Flax was a herbaceous plant that would be left to rot, then dried out so that the valuable fibrous material could be obtained.[56] It was then made into linen, which at that time was what the poor had as an equivalent to the silk that the wealthy, for instance, had incorporated into their clothing. And the use of linen can be dated to the arrival of the horizontal loom and spinning wheel in England around the late medieval period.[57] The materials that this instrument produced, thus, also had the effect of expanding the role of the bone collector, who now became a rag-*and*-bone collector. But this was a logical development since it occurred at a time when everything that had, in one way or another, been raised up out of the earth was in turn used and reused until its value or potential had been realized. Depending on the material, it would probably also be returned to the earth if no further use could be found, in order to avoid 'crises of subsistence'.[58]

The beginnings of the decline of the rag-and-bone collector in the twentieth century can be indicated with reference to two stories from Manchester, England, separated by less than a decade. They give some indication of the unpredictability of what was, in large part, a seasonal trade to begin with – dependent on weather conditions – if not a line of work in which the luck of the find was a key to economic survival for the collector. In 1951 the trade in Manchester was said to be booming, with the prices that could be fetched by rag-and-bone collectors, when they took their haul to a dealer, driven by shortages of materials such as woollen rags and cloth. These had significant value as materials that went into the production of new woollens and, especially, paper.[59] At the same time, other discards, especially clothing rags, were becoming harder to obtain in the thrifty post-Second World War era as people were 'wearing clothes longer' and keeping material scraps that otherwise might have been thrown away, for patching the clothes that they did keep hold of and wear. In Britain, clothes rationing – as part of which the government made films showing people how to mend clothes – had only ended in 1949.[60]

The situation meant that the rag-and-bone collectors had to work harder to find – among what one report described as a baffling '750 grades of rags' – the real items of value. 'I can tell them in the dark,' one collector said, 'but you can still lose out on a deal if you aren't awake.'[61] But in the act of finding in itself, the work of realizing the value of the rags was far from done: 'Then you have to clean them – take off all the linings and trimmings and buttons – and sort them out into cotton rags, whites, merinos, and so on. [The haul] looks a bit smaller when you have finished.'[62]

By 1958 it seemed to be more of a struggle for the collectors, who took out and returned carts to large traders who bought their rags, bones and other goods, including furniture and other objects (of steel, glass and metal). It was not unusual for one of these collectors to cover 20 miles, heading out from the centre of Manchester to far-flung suburbs, dragging a cart from street to street. On the newer concrete roads, it could be relatively easy going, but 'on the loose, nonslip surfaces and the setts it seemed heavy, even when unladen' – and all to offer the housewife rubbing stones (used for scrubbing doorsteps) in exchange for unwanted items. 'You could earn a pile one time, but not now,' one rag-and-bone man told a reporter.[63] At the same time, attempts to get at the truth or otherwise of the riches that could be made by some collectors usually came up against silence. Those who found themselves being asked about how much they earned had an interest in avoiding going into too much detail. There was always the danger that the public whose unwanted stuff they took away in exchange for rubbing stones or such items as balloons, cheap plastic toys and goldfish might think they were being conned.[64] By the final decades of the twentieth century, the informal methods of disposal provided by rag-and-bone collectors had largely vanished from the towns and cities of Britain, and the work they did would, in part, become the domain of technical specialists and located within impersonal systems of waste management, which, of course, included the infrastructure that developed around post-1970s 'recycling'.

The second-hand economy in the 1970s was understood to be part of a wider and growing interest in the ephemera of times past (a trend whose broader contours have also shaped understandings of the contemporaneous appearance of heritage culture and postmodernism, in which the notion of 'recycling' takes on a more fluid meaning). 'Flea markets are booming,' the *Guardian* newspaper reported in 1976:

Salvage yard in Willesden, London, 2016.

> Carried forward on a wave of nostalgia they are colourful proof that everything has its price and a buyer . . . On sale seems to be everything from sets of cigarette cards to flappers' dresses to spare parts for 1930s cars. The enthusiast and the collector rub shoulders with those who just stare and mumble why did they ever throw anything away.[65]

In London, the revivalist fashion boutique Let It Rock, opened by Malcolm McLaren and Vivienne Westwood in 1971, drew the attention of America's *Rolling Stone* magazine, which declared it the harbinger of a new interest in 1950s clothing and other 1950s ephemera (including old valve radios), drawing in customers from all over Britain.[66] It built its stock up from unwanted clothes 'sourced in markets and from suppliers who had leftover 1950s stock locked up in warehouses'.[67] The popularity of the film *American Graffiti* – released in the summer of 1973 – stimulated its own wave of interest in 1950s and '60s American clothing, often sold in stores that specialized in, and sold only, 'Americana'. Other fashion trends were also inspired by film: an interest in 1930s clothes sparked by *Cabaret* (1972), for instance, and the 1920s look of *The Great Gatsby* (1974), were catered to in London fashion stores like Biba with newly made retro styles (if no second-hand versions or approximations could be found).[68] In time, this way of 'recycling' would be theorized as symbolic of a growing wave of 'retromania' that swept popular culture in the 1970s and '80s as the postmodern era began to become evident in various kinds of temporal disjunction, such as the mixing of old and new, past and present, in forms of dress, which had first been found in the London boutiques of the late 1960s 'that merged fashion and antiquarianism, the psychedelic and the passé'.[69] In the words of a 1987 newspaper article reflecting on the upending of the world of designer modernism and high street fashions, 'the signs of retroism are everywhere.'[70]

But more broadly than these trends, second-hand clothing, as well as unwanted or discarded old (out of fashion) stock, was a staple of the London youth fashions of the late twentieth century, in particular.[71] It appealed to those who were after something unique that might help distinguish them from the crowd. Used items of clothing were often one-offs and, as such, an item like a rare silk scarf or a hat could add a highly distinctive touch to one's personal look. So, when it came to styles of dress, there was still an element of the fashion follower's desire to develop a unique look, but finding in the search for the new – as it were – something in the old and discarded. The roots of a certain hippie style in 1960s London were also found in a thrifty ethos and the possibilities that discarded clothes from earlier eras offered to those who preferred to 'create a unique appearance out of a bricolage of already used clothes'. But there was often more to it than making a fashion gesture, for one's appearance served a double purpose 'against the wastefulness of the consumer society', as Elizabeth Wilson has noted: 'You rejected the mass-produced road, and simultaneously wasteful luxury, and produced your own completely original look.'[72]

Stellar Deals without the Landfill

In the decades that followed the emergence of the new concept of recycling, there were also many adaptations of the general idea. These were always related to ways of extending the life of objects and products or finding new uses for leftover materials – upcycling, freecycling – sometimes facilitated through early Internet exchanges or online classified ad websites like Gumtree (UK) and Craigslist (USA), all of which could have been said to extend the circulation of second-hand goods and materials. By the early 1990s, I was working on a stall at weekends in Glasgow's massive flea market, the Barras, which sold literally anything and everything you could imagine and much that you would never have thought existed. For around four years I occupied an outdoor covered pitch with display tables that were erected at the start of every trading day, no matter what the weather. Right next to me was a patch of waste ground on which another trader, unable or unwilling to pay for a better pitch on which a covered stall could be erected, would spread out whatever miscellaneous junk he had ready to hand that day. For my part, I sold fabric that either contained very slight imperfections or came as leftovers from much larger rolls of fabric (of insufficient yardage to sell through normal retail channels), and sometimes discontinued designs that the manufacturer had no other way of getting rid of than throwing them away at a total loss.

All of which is to say that aside from the sorting of plastics, glass, cans, paper and other materials, much of what we think of in terms of 'recycling' was, until relatively recently, just absorbed into the second-hand economy. As the Internet became more accessible in the 2010s through the prevalence of mobile technologies and always-on networks, platforms like eBay, which had started essentially as a virtual junk store (and today occasionally markets itself as a way of satisfying consumer needs while saving unwanted goods from going to landfill), also helped to create wider acceptance for online digital payments, its partnering with PayPal serving to normalize such transactions as just another aspect of daily life. The success of the latter fed an increase in online buying and selling, expanding trust in the new impersonal forms of transaction and helping to make second-hand buying and selling something that anyone connected to the Internet could take part in. As a result, the idea of a virtual marketplace in any and every kind of object and product, no matter its provenance, quality or legal status, was born. eBay, the most prominent of these, is a gargantuan global marketplace but one that expanded out from its initial virtual flea market origins, first providing an

Swap meets, flea markets, car boot sales and similar places and spaces are sites for the transformation of waste, Los Angeles, California, 2018.

outlet for all those goods that were unwanted (but not without value or beyond use), then expanding further into a place where private sellers and, indeed, companies of all sizes now sell brand-new goods as well. Nevertheless, it remains today a means through which many earlier forms of waste-saving and restitution or reuse are able to continue, keeping material goods in circulation after they have fallen out of use: broken or malfunctioning electrical equipment and appliances, car parts previously located in scrap-yards, imperfect or second-hand clothes, and untold varieties of junk and ephemera (the non-waste environmental cost of that continued circulation, in terms of energy use, emissions and other factors, is another matter altogether).

But when it comes to flea market selling and buying, there are also significant differences between the real and the virtual, as cyberpunk pioneer William Gibson explained when detailing his own eBay obsession. Browsing the near infinite number of items for sale, he claimed, seemed to activate a 'hunter-gatherer' impulse, with the pleasure of search and find only intensified by what he described as the 'modes of sheer drift' that carried him through a virtual market that seemed without limit:

> Every item offers you a chance to peruse Seller's Other Auctions, which can take you off into categories of merchandise you wouldn't have thought of. A search for Hopi silver, for instance, brought up other kinds of Native American artifacts, much older ones, so that a series of clicks through stone adzes and Clovis points led to an obscure monograph on mound-excavation in Florida in the 1930s.[73]

This provides a neat example and summation of the riches to be found – by seller and buyer alike – in the global opening-up of those storage closets, lofts and garages that once provided a resting place for things that had fallen out of favour or no longer conformed to the fashion of the day. One could continue to detail examples of how the recycling principle has expanded into more aspects of everyday life in contemporary society. For the French philosopher Jean Baudrillard, for instance, postmodern culture represents a distinct historical epoch of near universal cultural recycling. Yes, he admits, it might all be related to the fact that everywhere today we are persuaded to recycle actual physical and material waste, but there is more to the idea of recycling now.[74]

The twin forces and excesses of cultural production and cultural memory – including the imagination, and habits and practices that become unmoored from the time and place that gave them meaning in the first place – give rise to countless reinventions and resurrections. Disneyland, Baudrillard notes, was one model of this ideal, but one in which recycling takes us back to waste in another sense; a place in which 'the dreams, the phantasms, the historical, fairytale legendary imaginings of children and adults alike is a waste product, the first great toxic excrement of a hyperreal civilization.'[75]

What remains strange about the basic or original application of the concept of recycling, if we understand it to refer to the indeterminate state of something that was discarded and awaits – in a sense, reanimation – is how uninteresting the sphere of waste that it refers to is. The things that enter the recycling stream disappear and make waste disappear. Within the scope of the broad approach to thinking about the human relationship to waste in its various forms or modalities, around which this book is organized, recycling, as both a concept and a phenomenon of the consumer society (distinct from all the other practices that are unthinkingly redescribed as 'recycling', and which are much more culturally engaging and fascinating), is something that has not been as readily amenable to the cultural imagination in the way that certain other persistent and visible phenomena of waste are. The object to be recycled, per se – for example product containers made of plastic, glass, cardboard and so on – is only temporarily waste and as such exists as a material thing awaiting the processing that we hope transforms it either back into the primary material element it was first constituted from or into something else. As a thing to be treated in this way the object to be recycled, which has already been moved out of the sphere of waste, is not generally claimed by the creative imagination, but by a technocratic logic that is concentrated on issues to do with tracking, measuring and managing materials and resources. Unlike other modalities of waste – one thinks of trash mountains, sewers and deserts, and even the urban flea market – it is not that easy to represent something that is stripped of the qualities that allowed it to serve, for instance, as container or packaging, now that it is reduced back to some primary material form. The object to be recycled does not linger in the human realm or indeed the imagination for long, if at all. It exists simply in the space between being disposed of and being carried away by others. And beyond a civic duty of depositing recyclables in a place where they can be collected, we have no further material or significant part to play in the reanimation of such wastes or their reappearance.

This is perhaps why recycling itself is so opaque, and why today, several decades after the idea first gained currency, we see it claimed that recycling is broken or it doesn't work.[76] Whether or not such claims are accurate would seem to be a matter for an economist or some other analyst who would fall back on catalogues of figures, quantities and other forms of information which, for most laypeople, would be difficult to make sense of.

Globalizing Waste

That the stuff of waste becomes something spectral, or something that vanishes into places and systems or relationships that we have no interest in, was a subject amusingly dealt with in the popular TV drama *The Sopranos* (1999–2007), well known for its depiction of a Mafia family who cultivate an ironic self-image among themselves that has been infused with elements of representations of the Mob from popular culture, such as Francis Ford Coppola's *The Godfather* (1972). In one scene early in the series we join Mob boss Tony Soprano as he takes his daughter, Meadow, on a tour of some of the elite East Coast universities that she would like to attend. At some point during a trip that takes them away from the rest of the family and symbolically lifts them out of the everyday concerns that help to ensure that what is familiar in everyday life remains unremarkable, she suddenly realizes that they – her family – don't seem to be like the other people who occupy this world. Her father clearly doesn't belong at all to this world of colleges and professors and seems unable to interact with the people he encounters in a normal way (the viewer almost expects him to offer bribes or threats to ensure his daughter gets into the college of her choice). It raises in Meadow's mind the question of what exactly it is that he does for a living: 'Dad . . . are you in the Mafia?' she starts to ask. 'What?! Who told you that?' '. . . or organized crime, or whatever you call it?' 'I'm in the *waste management business*,' Tony responds emphatically, trying to dodge the issue. 'Everyone automatically assumes you're mobbed up.'[77]

Most of the time, though, we don't see Tony and his associates doing much waste management (aside from the odd scene where we glimpse them illicitly tipping the contents of a garbage truck into a nearby bay in the dead of night). While those kinds of depictions amuse us, they once again point to the fact that, for the most part, we have little idea of what happens to our wastes after they are taken away.

Don DeLillo's *Underworld*, discussed earlier and probably the greatest and most thoroughgoing vision of waste in all its manifestations in twentieth-century American life, carries a title that may also be taken as an acknowledgement of the mysteriousness of the science of waste disposal and management as it had developed by the end of the twentieth century. Like *The Sopranos*, it lends itself to the idea that the world of waste management had – at least in the eyes of popular culture – become the preserve of gangsters, if not a global market that made the stuff we cast aside disappear in new and more nefarious ways. The title 'Underworld', one observer has noted, 'humorously recalls the Mafia and jokes associated with it', because in New York, DeLillo's home town, 'waste management was one of the industries most often connected with the mob.'[78] Indeed, a recurring image of waste in the book is of a 'spectral ship', in search of somewhere to land its suspect goods, possibly owned by the Mob and thought to be filled with hazardous materials. DeLillo's waste-obsessed characters speak of the stories they have heard of it drifting for years around the coasts of various locations, unable to dispose of its cargo – the Hudson River, Liberia, the Philippines. Nick Shay thinks it is a drug shipment disguised as toxic waste, but his colleague Sims is not so sure: 'One, it's a heroin shipment, which makes no sense. Two, it's incinerator ash from the New York area. Industrial grade mainly. Twenty million pounds. Arsenic, copper, lead, mercury.'[79]

DeLillo might have based his spectral ship on the case of the cargo ship *Khian Sea*, which was headline news in the United States in the mid-to-late 1980s as it roamed the seas from the east coast of the country, around South America and across the Pacific Ocean to Asia, looking for a place to dump its load of toxic ash, now all part of the growing global trade in dangerous waste products. As Kathlyn Gay has written, much of this waste – sent by wealthy nations in the West and northern hemisphere to poorer regions of the world, in the East and southern hemisphere – was misleadingly described as 'recyclable' in order to get rid of it: 'a common way to dispose of' those wastes most difficult to process.[80] Such materials would include carcinogens, chemicals, nuclear waste, medical waste, infectious materials, mercury, explosives – all of which are by-products of the comfortable lives people enjoy in the wealthier parts of the world. Gay notes that the industries responsible for the manufacture of 'chemicals, plastics, pesticides, computer parts, prescription drugs' and other materials were also therefore 'responsible for releasing a great variety of toxic chemical wastes into the soil, water, or air'.[81]

In the era of globalization, a sometimes regulated, sometimes obscure movement in waste between the global North and South and the West and East developed as a shadow of the exchange of goods that characterizes the global consumer economy. The situation where waste in one place represents value in another reflects the fluid or shifting nature of other kinds of waste, which fluctuate between usefulness and worthlessness according to the economic status of those involved in the exchange. The new ecological awareness of the post-1970 era, marked by the first Earth Day in April of that year, eventually changed what happened to waste, although perhaps not in the ways it had intended. In the United States, more recycling versus older forms of disposal meant newer players in a waste industry where 'strong-arm' business tactics were not unknown.[82] 'Technological advances and environmental consciousness changed the way waste was disposed of,' wrote Harold Crooks in his 1993 book *Giants of Garbage*, thereby 'making both disposal facilities and waste increasingly valuable commodities'.[83] What followed was 'fierce competition across North America and then farther afield for control of disposal sites and waste flows'.[84]

The idea of the world beyond the borders of any given nation becoming a potential site for dumping waste opened up a new space for the excesses of the rich countries of the West and North, who were now subject to more stringent environmental regulation. In using the developing economies of China, India and the nations of Africa as dumping grounds, they mimicked the logic of waste management that had always been the default: out of sight, out of mind. That is not to say that it is fair on those nations that find themselves in a position where there is some economic benefit to absorbing, salvaging or recycling the garbage of the rich. For almost three decades between the early 1990s and 2018, China – to choose one example – was in need of raw materials, predominantly plastics and paper, to drive economic growth, and, as a result, accepted large volumes of waste for recycling from the West, but often in the 'most contaminated, least valuable' varieties.[85] This was due in part to the failure of those recycling the wastes at their point of origin to separate (and clean) materials to be recycled from non-recyclable materials and food residues. The result was that much of the plastic waste sent from the United States to China was never recycled, instead being dumped or burned 'outside China's impoverished, unhealthy "recycling villages," the shantytowns full of mom-and-pop recycling businesses that lined the edges of China's big port cities, reeking from caustic chemicals and burning garbage'.[86] That the rich world trades not only waste but additional forms of pollution, which result from the fact that these wastes are used or processed in ways that would not be permitted under the environmental controls that exist in the countries of the North and West, once again reveals that waste can easily multiply or proliferate in new ways as a consequence of allowing it to elude established practices or systems through which it is controlled.

Some material moves across the globe in rather different ways – nuclear waste, for example, is subject to 'reprocessing'. At the Sellafield site in northwest England, for instance, spent fuel from around the world was 'recycled' into new fuel for several decades between the 1990s and 2010s at the now decommissioned Thermal Oxide Reprocessing Plant (THORP). Other 'high-level' waste continues to undergo vitrification ('dried to a powder and mixed with glass at a temperature of around 1,200 degrees Celsius', according to Sellafield).[87] Low-level waste (contaminated 'gloves, protective clothing, paper towels, metal and concrete') is encapsulated in concrete and then, in the words of one recent anthropological study, made into new 'wasteforms' as part of a process of packaging and repackaging: 'secured in boxes, drums, containers, flasks'.[88] These objects are new forms of waste because they still need to be disposed of in a long-term geological disposal facility (GDF), which will be required to contain them for a period of at least 100,000 years, and which at this point does not exist.[89]

4

Aesthetics: Designing and Dematerializing

Smooth bodies in muted shiny colours seem to glide past as we look on; chrome highlights, grilles and wheel spokes loom into view. Tailfins, chamfered edges and glass that sparkles like a rare jewel jut out of the face of this sculpture of modern desire, mass produced and driven off the production lines of the factories that once powered the economy of what came to be known later as the American Rust Belt. Here the object of desire is in Barry Levinson's 1987 film *Tin Men* – an evocation of a culture at the apex of its consumerist triumph – rotating on a raised plinth, the newest and most alluring Cadillac so far. 'Isn't she a beauty?' says a salesman. Set in 1963, *Tin Men* reveals the ways in which obsolescence depended on the affective relationship people had with the object world, an almost elemental relationship to what was newer, brighter, better than what the neighbours had.

Surplus Aesthetics

If the second half of the twentieth century represented an acceleration in living, it implied that we had entered an era in which the world was always becoming different from itself. The temporality of fashion had been internalized. In the design of many objects that would come to exemplify the idea of affluence that permeated life, there was, as the critic Reyner Banham wrote in 1955, a

Cars, as much as any other product in the decades following the Second World War, exemplified how the desirability of new forms and designs was bound up with obsolescence.

developing aesthetics of expendability at work – in things like the design and marketing of cars, but also in the culture more widely, touching almost all spheres of life. For 'expendability' as analysed by Banham, we may read what in the post-*Waste Makers* decades would be described as obsolescence. The sleek presence of the new car in one's life, or the new kitchen, was a thrill that was destined to become eclipsed by something that possessed greater power, an object of desire that made the known or familiar suddenly old before its functional life was even close to being over. 'Aesthetics' in this sense refers not to the study of beauty (as it did in art of the eighteenth and nineteenth centuries) but is rather akin to the original meaning of the word in being connected to the realm of sensation and feeling (although, as with the salesman's pitch in *Tin Man*, we can see an appeal to the beauty of a mass-produced object of desire). The aesthetics of capitalist materialism, as a form of emotional engagement and attachment expressed through the world of goods and possessions, illustrated an aspect of what

Banham, writing some years before the term 'throwaway society' was coined, described as a 'throw-away economy' that reoriented consciousness, making people see their 'worldly possessions' in terms of how expendable they were.[1]

At the same time, the objects of the new popular culture, Banham noted, had to be desirable and novel. But this expendability was simply, on a smaller and personal level, an aspect of a broader cultural trend within which the designed world, and the built environment, took shape. Expendability was evident everywhere but varied within the normal – or normalized – expectations of how long any given designed object or structure would be likely to last. Architects might create structures that could last a very long time. But the contents of a building and the uses its interior design permitted (also a possible factor in determining how useful a building was over time) would likely have a shorter lifespan and become expendable or obsolete much sooner than the structure itself. The contrast between these two apparently disjunctive temporalities is seen in contemporary trends of retrofitting, which retains the main structural elements of a building but adapts its interior to new uses. But the culture of the expendable designed world was, according to Banham, also built into the human desire for improvement and transcendence at a more fundamental level: the lure of a future that might be different, or might represent an upgrade on the present, had become a defining feature of life. It went beyond consumer culture and into the general intellectual orientation of the times.

The accelerating world led to new discoveries, new technologies, that also made expendability part of the fabric of a new reality. In this climate, Banham observed, even those things that took up no space, such as mathematical models, or very 'valuable' creations like exploratory spacecraft are highly expendable, serving specific purposes that may be realized in a short amount of time before they are dispensed with.[2] The point was that by the mid-twentieth century and specifically in terms of culture, aesthetics and intellectual adaptation to new problems, Western societies were engaged in the continual adjustment of consciousness to keep up with a reality that was seen to be inherently unstable. In the earliest days of mass consumption, General Motors in the United States felt able to claim that their 'big job' was to make obsolescence an integral device in economic growth: 'In 1934 the average car ownership span was 5 years; now it is 2 years. When it is one year, we will have a perfect score.'[3] There was a kind of inevitable logic to this shortening of consumer cycles. As an idea it was taken to comically absurd lengths in a 1970s British TV series, *The Fall and Rise of Reginald Perrin*, in which a small shop called 'Grot' that sells useless products ('Only the Best Rubbish') grows into a successful business empire. We see Perrin interviewed on TV about how he came upon this idea. 'It's really not such an extraordinary idea,' he declares, as:

> Most of our economy is based upon built-in obsolescence – I just build it a bit further in. The things are obsolete before you even buy them. I haven't gone as far as I'd like to. Ideally, I'd like to sell things that fall to pieces before they even leave the shop. What a gift to capitalism that would be! 'Oh – it's fallen to pieces. I'll have another one.' 'Certainly sir.' 'Oh – that's fallen to pieces too. I'll have another one.'[4]

But if hastening obsolescence seemed like an inevitable consequence of the way the economy functioned, a simple outcome of the acceleration of life in advanced societies, in which modernist design ideals had become commonplace, then instantaneity, novelty and disposability also became the chief virtues of the age. They would be exploited for their ability to drive capitalist production to ever-greater heights of turnover but tending always towards the production and consumption of a material reality that, eventually, took shape as a shadow world of waste. For some, such as the German Marxist critic Herbert Marcuse, who was then living in the United States, the fact of waste was – for the most part – concealed not just by the way it was rationalized out of life, but by a powerful illusion that shaped consciousness: the idea that freedom of choice was self-fulfilment, or self-realization, attained through the world of consumer goods which, in Marcuse's words, existed as the numerous competing 'brands and gadgets' that one had to choose between.[5]

Following Banham, design historians have explored why 'waste became an essential component of the production cycle.'[6] The simple answer was that market demand had to be stimulated and managed to condition the appropriate response, which was a constant need for new products. Otherwise – if the consumer held on to products for longer – there would soon come a point of saturation that dampened demand and ultimately threatened prospects for continued economic growth. Hence the logic of planned obsolescence went hand in hand with the belief that people must be conditioned to buy things, 'not once but several times, thereby stabilizing the production cycle'.[7]

The smartphone of today provides an interesting example of how the tension between consumption and disposal is maintained as a means of stimulating and growing a consumer market that can be led back to the stores for newer, improved versions of their devices, with the relentless schedule of software upgrade cycles inevitably followed soon after by the need for new hardware. The idea of built-in obsolescence here vanishes to a great extent under the clamour for better or more finely tuned performance, new functionality and so on, which in the end necessitate the purchase of a new product, if one is to keep up with the pace of technological change as it affects everyday experience. Such products, to put it another way, come to be viewed as imperfect – indeed superfluous – regardless of their material condition. The materiality of the object in this case is of secondary importance because it is not an object in a conventional sense, but an interface to experiences and services accessed via networks. The objects are then susceptible to being regularly updated or reconfigured in some way on 'the other side' of the user/service divide until a device is no longer able to function optimally. With devices such as the Apple iPhone, seven years is the maximum length of user support (that is to say, the length of the time, from new, that the device will be able to receive upgrades that allow it to provide an up-to-date experience), although as many stories in recent years have revealed, in practice, the period of usability of these devices is often shorter than this. This is, of course, just a newer and more ingenious form of 'built-in' obsolescence, in which the products are eventually rendered useless because they cannot keep up with the pace of change in the real world.

Frank Gehry

If artists could be said to have developed a range of distinctive attitudes and approaches to the excesses and leftovers of modern life, so too did others, working on a different scale and towards their own sometimes unspecified or unknown ends. In the designs and creations of Canadian American architect Frank Gehry, we see maverick attempts to bring waste out into the open as both a resource and as an element of a new aesthetic that stands against the expendability previously identified by Reyner Banham. After turning to architecture relatively late, by the 1960s Gehry had found himself running a small but successful architectural practice, although he was often engaged on projects that didn't really offer the creative outlet he was looking for. One day when he was putting together some scale models from cardboard, a preferred method of quickly testing a design idea, he realized that this material – so often seen as useful only for packaging things of value and not really of any other value in itself – made something very strong when it was glued together in layers. And so, from this typically throwaway packaging material he began to see new possibilities for furthering the use of cardboard, transforming it into novel forms. He began to design cardboard stools and chairs, which would eventually develop into an inexpensive range of furniture known as 'Easy Edges' that was sold in Bloomingdale's stores at prices ranging from $15 to $115.[8] 'The cardboard', noted one retrospective on Gehry's work, 'projected the wholesome connotations of corduroy or wood, and was in fact stained, joined and generally worked as wood. Three of the chairs could even hold up a car, as one of the publicity shots demonstrated.'[9]

But after the first line of cardboard furniture was sold out, Gehry decided that as interesting as it was as an experiment, his main focus would have to be architecture. He had, nonetheless, now established the principle of using materials of negligible value for something other than what they had been intended. If the cardboard furniture brought him to wider public attention for the first time because of the novelty of his attempt to bring such cheapskate objects into the homes of ordinary Americans, it was merely a prelude to the fame he would acquire as a result of the redesign of another domestic setting – his own home.

As the historian Kevin Starr wrote, Gehry shared with Simon Rodia, creator of Los Angeles's famous Watts Towers (a structure built with junk materials), a desire to reveal 'the beauty inherent in industrial materials' when brought into relation with organic forms of the surrounding environment.[10] The most famous demonstration of how Gehry tried to rethink the relationship between space, form and material, was – up to that point – his project to remodel his own residence in Santa Monica, Los Angeles County, in the mid-1970s. This was a house 'deconstructed' from the inside out, using materials that were perceived to have no place inside a family home, to fashion a new interior, and which left the frame of the old house as little more than 'a kind of scaffolding memory' that could be seen surfacing or peeking out from certain vantage points.[11] It was as if junk had in some way begun to organically grow over the frame of the house, much in the way that trees, moss and other plant life grow out of abandoned concrete buildings.

Gehry's house baffled and outraged his neighbours, but the attention it gained in the press inadvertently marked a key turning point in his career. To the casual observer obtaining only the most fleeting or superficial of impressions from the street-facing facade, what was termed by one observer a 'shipwreck of a house' might have seemed to be a construction that was part wild beach shack, part house, which is to say made from materials that looked as if they had been salvaged from the leftovers of a storm that had blown in from the Pacific.[12] There was corrugated metal siding,

Frank Gehry, 'Easy Edges' chair, 1970.

plywood, chain-link fencing and other unlikely materials, all of which had been reconfigured in the most unusual way, sprouting at odd angles to form a protective shell around what in fact had originally been a modest pink bungalow of the kind said to fulfil dreams of domestic comfort. 'The stark effect of the corrugated metal frame', the critic Fredric Jameson wrote, 'seems to ruthlessly cut across the older house and brutally stamp the mark and sign of "modern art" on it, yet without wholly dissolving it, as though the peremptory gesture of "art" had been interrupted and abandoned in mid-process.'[13]

It had, indeed, the look of something unfinished or perhaps thrown together as a temporary measure. It also seemed as if the house were a rebuke to the neighbourhood that it sat in, giving the appearance of having been turned back to front – a flash of the backside – to present the ugliest-looking surface to the outside world. But, on the other hand, perhaps it was all a matter of locational context and this apparent garbage house had to be seen within the context of Gehry's other Los Angeles buildings. If one took the view of Mike Davis, one of Gehry's critics, it could be taken as another example of what he described as the architect's 'stealth houses', so-called because they concealed 'their luxurious qualities with proletarian or gangster facades'.[14] High walls, barbed wire and barred windows, the hint of armed security lying in wait.

At the time of the house's first public unveiling, the *Los Angeles Times* reported that Gehry's neighbours were dismayed by the sight of what appeared to be a hotchpotch of out of place materials that had suddenly come to occupy a prominent position on their street. The house, some said, was 'a monstrosity', 'anti-social', a baffling structure that looked more like a 'prison'.[15] But others seemed to recognize something inevitable in the fact that such a building had ended up in that spot – in Santa Monica, a short distance from the ocean – as if it were simply another demonstration of the dictum that if someone were to pick up the USA and give it a shake, anything not fixed down firmly would naturally end up in Los Angeles, and by extension continue rolling and tumbling out to the edge of the land. 'This is Southern California,' a more understanding and informed neighbour said, 'one expects all the loose nuts and bolts to fall. Sure enough, we collect 'em.'[16] It wasn't clear whether this reference to 'nuts' was aimed at Gehry or to the fact that Los Angeles had long been perceived to celebrate trash culture.

As well as Gehry's incorporation of what looked like leftovers from a construction site, there were the materials that one normally did not associate with houses in well-maintained residential districts, but which the architect – upending conventional aesthetic norms – regarded as something akin to beautiful sculpture: 'A cage made of cyclone fencing. Some artists like to work in oils, some in marble, and Gehry likes to work in cyclone fence, which he has incorporated as a design feature in a number of his projects.'[17]

Against the critics who attacked his use of this particularly despised material, Gehry would say that he used it not out of any love for the stuff, but simply because it was so abundantly produced. 'I'm horror struck at the stuff,' he claimed, insisting that he was merely reusing what would otherwise go to waste: 'if the culture makes all this baloney, why not use it?'[18] Inside the house, meanwhile, there was also an attempt to show how it was put together, visible in details such as the stripping away of materials to reveal structural elements. Beams and joists that had previously supported an upper floor before Gehry acquired the house were now removed to reveal details that were normally hidden, another instance of turning things inside out. A roof created by a tilted cube of glass opened the kitchen out to the sky, an example on a small scale of the disorienting features he would later become famous for, and which Gehry thought gave the house

Frank Gehry, Gehry Residence, Santa Monica, Los Angeles County, 1977–8.

the feel of a vessel that had been partly submerged in the ocean. Another smaller window, he proudly declared in an essay for the journal *Design Quarterly*, was improvised with a hammer, which was weaponized to make an opening in a wall over which he then glued a pane of glass, the entire feature a piece of hack design. 'It never leaks,' Gehry said of this most rudimentary of windows. 'All the others do.'[19] There was little doubt that aside from the critics and doubters, Gehry had started 'a revolution of sorts' in bringing to the world a form of 'deconstructed' architecture: 'plywood walls for interior spaces, corrugated iron and chain-link fencing for exteriors that celebrated metropolitan Los Angeles as an unfinished city'.[20]

If Gehry's house represented a revolution, it was one that took place at the rarefied level of architectural design and artistic expression. Within the context of the decades that followed, when Gehry became known for projects much larger in scale, his remodelling of the small 1920s wooden bungalow constituted a more free-range, anarchic approach to the possibilities of domestic architecture. Much like his experiments with cardboard, the Santa

Monica house had allowed him to break out of a period characterized by uninspiring commissions for more routine spaces, such as shopping malls and offices. The possibilities of operating at what would be considered a more artistic level of creativity had long occupied his thoughts. The success of this relatively small and modest personal project was taken as a sign that there was a way to make a living from creating less predictable architecture. It led ultimately to the huge projects that Gehry delivered in the following decades, such as the Walt Disney Concert Hall in Los Angeles and the Guggenheim Museum in Bilbao, structures unlike anything seen before, as well as others that one critic dismissed as showy postmodern gestures, like 'fixing real aeroplanes to the side of buildings and designing others so they looked like fish'.[21]

Garbage for Leisure

In the United States in the second half of the twentieth century, right around the time when the excesses of consumer society and the awareness of how imperceptibly life had begun to accelerate were running into the first wave of environmental consciousness, one response was to utilize the increased volumes of waste that orbited cities on its way to various dump sites as the stuff on which new leisure and landscape parks might be built. Interestingly, such wastes would be taken out of the cities and used as a novel method for making use of what would otherwise be regarded as wastelands; sites for which there was no other obvious or productive use. And so, rather than being deposited in landfill, as most urban waste material was by then, it would be used as the material from which new designed landscapes would be sculpted, and which often took the form of mini mountains, ski slopes and other pleasure grounds. Rubbish collected from cities served, under these methods, a dual purpose in that it could be reshaped into any form to outline the contours of a landscape (and vanish from view once covered over with grass or other plants), and at the same time allow city planners to reclaim land that may otherwise have been unusable, much in the way that common land (also known as waste or wasteland) had been made productive by transforming it in some way over the previous four or five centuries. The creation of beautified garbage landscapes, however, can be seen as another by-product of the throwaway society and its excess material consumption: a case of the leisure society creating more leisure spaces.

At the Roy C. Blackwell Forest Preserve in Illinois in 1967, a University of Chicago academic by the name of John R. Schaeffer was overseeing the construction of one of the first so-called garbage mountains – the very first one, constructed at this location, was 57 metres (188 ft) in height – to be set within a manufactured scenic landscape, which, in this instance, was also to feature a large lake at the foot of the artificial mountain. The lake was to be dug out from a former gravel pit on a site that in total would cover nearly 16 hectares or 39 acres (or, to put it another way, a water surface that would be roughly equivalent to 624 tennis courts). At the time, the main means of disposing of city waste was in landfill, a practice first developed in the United States in the 1920s. The basic principle of landfill, as Martin Melosi writes in *Garbage in the Cities*, was to contain waste, while ensuring that organic materials that formed part of it didn't rot:

> Typical sanitary fills were layered: twelve inches of garbage were covered with eighteen to twenty-four inches of ashes, street sweepings, or rubbish; then another layer of garbage; and so forth. Chemicals were sometimes sprayed on the fill to retard putrefaction.[22]

And while earlier makeshift landfill techniques – through the ploughing of garbage into agricultural lands, for instance – began in the United States at the beginning of the twentieth century and pre-dated the sanitary landfill, the use of city wastes (within the American context) as material 'to fill ravines or to level roads had always been regarded as highly objectionable'.[23] It wasn't until the 1960s that 'filling low places with refuse or reclaiming marshland and coastal land became desirable for many cities.'[24]

One case being made for Schaeffer's idea of turning waste into the stuff out of which new leisure resorts would rise, aside from the main one of turning something unwanted into something more acceptable to the eye of the casual observer (who might never know what lurked underneath the surface), was that there was thought to be less risk of pollution from a city's waste if it was piled up overground and made into dry ski slopes rather than being buried, where leakages might contaminate land and seep into water sources. In Schaeffer's method, however, although the

Northala Fields, Ealing, London. A landscape of several fishing lakes and four artificial hills, created with waste material recovered from the demolition of the old Wembley Stadium and construction of Westfield London shopping centre, which opened in 2008.

garbage rose up during construction and was not buried in the way that waste deposited in landfill was, it nonetheless followed that example, utilizing a thick layer of clay to compress the growing mound and prevent 'any unpleasant smells from escaping'.[25] As an idea it seemed to have some appeal, seeing as there would be an inexhaustible supply of domestic and commercial waste for as long as there were humans and cities, and production and consumption. 'Of course when the mountain is completed, there will still be more garbage to get rid of,' Schaeffer told reporters, 'and I can't think of a better way of doing it than to build some more resorts.'[26]

And indeed, the principle was applied elsewhere. In Tyneside, in the northeast of England, the landscape that once belonged to Britain's National Coal Board and British Steel around the huge Consett steelworks (closed in 1980) – which once belched smoke into the air, sometimes leaving a film of red dust on houses and streets for miles around – was reclaimed to make a range of leisure sites for the pursuit of healthy outdoors living:

> an array of landscaped parks, clay pigeon shooting grounds, wooded plantations, football fields, nature reserves, paragliding clubs and dry ski slopes on colliery spoil-tips, travelers' sites, allotment gardens, all connected by a string of bicycle paths along the former industrial railway lines.[27]

The status of such sites in the eyes of developers, however, was sometimes indeterminate, with the leisure facilities seeming to act as temporary stand-ins to occupy a site until something more useful or profitable was found to do with the land. This idea of these spaces as mere holding sites would be characteristic of other post-industrial landscapes, which would be cycled through a succession of uses, but in this case the development was seen by some as another form of exploitation of the communities who had already been dispossessed by deindustrialization. These temporary leisure spaces represented 'a tendency', one critic wrote, 'to constitute intermittent collective places' in such a way as to reinvent industrial wastelands – depleted or exhausted and discontinuous with natural surroundings – as 'natural ground'.[28] In London, one of the most famous of the formerly toxic spoil heaps to be turned into dry ski slopes was found at Beckton, and could be reached on foot from east London by following what we could describe as a concealed waste route, 'the ridgeway of Joseph Bazalgette's Northern Outfall Sewer'.[29] In the words of Iain Sinclair, the top of the 'glorious absurdity' that was the now abandoned fake slopes of Beckton Alps provided a particularly interesting viewpoint on what he described as 'a panorama of blight and damage'.[30] By then, having exhausted whatever novelty it held when it first opened a decade or so earlier, Beckton Alps looked out over a scene that was far from the benign visions Schaeffer had when he came up with the idea of transforming garbage mountains into beautified designer landscapes. Below was a landscape littered with new eyesores, just a part of the continuing churn of urban entropy.

Designing Zero Waste

As previously noted, what distinguishes post-1970s recycling from the earlier forms of waste reclamation and restitution is the materiality that is specific to it. It involves plastics, glass, tin, paper, cardboard and other materials that arrive first as the packaging of the goods and products that we consume, and which are not easily or readily transformed for other uses without first being processed and then fed into the industrial production process.[31] The late 1980s saw the dawn of green consumerism, first appearing in Britain and continental Europe and encouraged by the publication of books such as John Elkington and Tom Burke's *The Green Capitalists* (1987) and Elkington and Julia Hailes's *The*

Green Consumer Guide (1988), and then spreading to other consumer societies, including the United States and beyond.[32] The trend was driven initially by concerns over the use of chlorofluorocarbons (CFCs) – greenhouse gases – in aerosols, awareness of which was raised at the same time as other global issues focused on the relationship between human beings and the natural environment began to make their way into political discourse.[33]

At the end of 1988, a list of looming ecological problems was detailed in one British newspaper, which declared that in the previous twelve months alone, 'awareness of our fragile environment has grown rapidly.'[34] A region within the Amazon rainforest 'about the size of Belgium', it was reported, 'went up in flames, billowing carbon dioxide into the atmosphere'.[35] But, in Britain, such events were also linked with concerns closer to home about pollution resulting from waste matter, with 'rubbish dumps at 1,300 sites all over the country revealed to be leaking potentially explosive methane gas'.[36] Don DeLillo's 1985 novel *White Noise*, in part about the spectre of airborne chemical waste and the uncertainties it produces in a population whose lives are greatly detached from knowledge of the dangers lurking in industrial processes, could be taken as a gauge of environmental concerns at the time, which were still driven largely by the threat to the environment from pollution.[37] DeLillo said that the novel was inspired by 'the frequent references to toxic spills in the news':

> The *New York Times* practically had a toxic spill page. On the television news in the evening you would have the sports, weather, and the toxic spill. But it wasn't until Bhopal that people even seemed to notice that this phenomenon was taking place. No one ever talked about these things.[38]

The authors of *The Green Capitalists* noted that incidents like the Bhopal disaster in 1984, which exposed hundreds of thousands of people to toxic gas, provided an opportunity for businesses to lead the way in taking action and setting an example by promoting an environmental consciousness. Concerns about damage to Earth's protective ozone layer, as well as a range of other issues, including 'the Greenhouse Effect, toxic waste shipments and the deaths of North Sea seals' had 'convinced ordinary people that to wait for new regulations may be to wait too long'.[39] The speed with which companies and manufacturers of all manner of products signed up to the new green agenda quickly led to the appearance of products sporting 'Ozone Friendly' labels, which allowed manufacturers to burnish their green credentials. It was a trend that was regarded with suspicion among some environmental activists, who decried any solution to ecological problems that was in any way consumer-based. At the conference of the UK's Green Party in 1989 – a year before it split into three regional parties – the first after the issue of the environment had exploded into the headlines, there was scepticism about the claims of supermarkets who now purported to be green. 'They do not give a damn what they sell,' one delegate told a reporter. 'The Green consumer to them is 15 percent more profit.'[40] Resisting calls to become leader of the Green Party, where it was thought he might be able to use the recognition he had gained as a TV presenter to drive the issues further into the mainstream of political discourse, David Icke – then one of the leading figures of the Green Party – added that 'supermarkets by their very nature are totally un-Green.'[41]

But if green consumerism and all its claims were seen as a bandwagon that producers and retailers were easily able to jump on to, there was another idea that soon emerged in what was essentially a battle to change people's consciousness by more directly raising the spectre of waste. That was the concept of 'zero waste', which can be dated to the beginning of this century. In a 2002 book titled *Zero Waste*, its author Robin Murray sought to advocate for its adoption as a means of shifting how society thought about waste. Our thinking, he suggested, had become lost in a dead end due to its focus on the problems of disposal. Discourses and imperatives that focused on waste minimization and recycling had, he thought, become exhausted. Zero waste was thus a concept, an idea, that could do more: it might provide a new means of galvanizing policy choices in the direction of more sustainable outcomes that brought together those aspects of disposal and reuse that were now well known, with responsible production, eco-design and 'smart' approaches to waste management.[42] 'Zero waste' might seem like a contradiction in terms, Murray added, or even an impossible ideal, because in truth there can be no possibility of eliminating waste entirely. 'If waste didn't exist we would have to invent it,' he noted, saying:

> Just as there can be no light without shadow, so useful matter, to have meaning, requires its opposite – useless waste . . . The test of commodities is whether they can become good waste. The problem of waste disposal is replaced by the problem of phasing out those materials which are hazardous and which cannot be recycled. The issue is not to get rid of them when they are finished but to avoid producing them in the first place.[43]

In 2004 Britain's Royal Society of Arts followed this trend in thinking with the publication of a manifesto for the new century that aimed to create a 'zero waste' society.[44] In that future society, design was to become a primary means of achieving waste reduction, defined in terms of how society could meet the 'needs of the present without compromising the ability of future generations to meet their own needs'.[45] Designers, in the words of a more recent author, would be 'the new alchemists, searching for ways to transmute existing resources into valuable new materials'.[46]

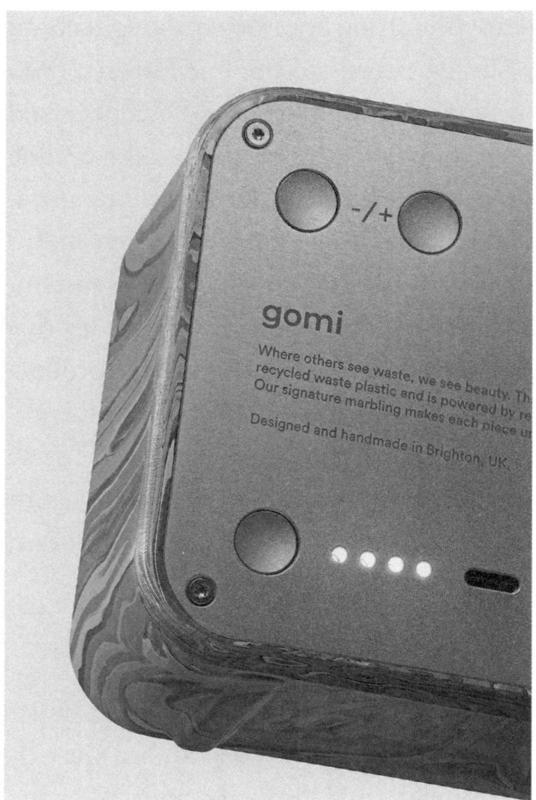

A speaker made by UK design studio gomi, based on a modular design approach that allows for the repair of parts that would otherwise be thrown away.

We can trace a line from green consumerism to the present era in which designers are – more than at any time in history – involved in developing products and forms of packaging that reverse principles of built-in obsolescence, aiming instead for 'zero waste'. At times it seems as if designers are generating ideas and solutions that would not be out of place in Callenbach's *Ecotopia.* Which is to say, they are designs that embody a principle of 'vanishing waste' (even if the vanishing is only temporary, or to be understood in one form of its materiality before it is reconstituted as another): wallboards made from the by-products of sugar refining; beer made using recycled sewage water; waste coffee grounds as biofuel or facial scrubs; 'compost cocktails' made from ingredients that would normally go in the bin; clothes made from plastic bottles; edible wrappers for yoghurt and ice cream, and so on. These represent efforts to upcycle and repurpose waste products. The idea is that there are always new sources of value to be found or practices to be learned (or perhaps relearned) to keep materials out of the waste stream for as long as possible.[47] The circular economy, as it is now known, extends the trend of the last fifty years or so, moving us towards behaviours that not only extend and embed the recycling of our unwanted stuff into new disposal routines and habits that work against the underlying logic of waste, but which are driven by waste as a shadowy presence that may loom into view unexpectedly if we ignore the work of keeping it at bay. When we recycle anything, it makes us use time in certain ways, and makes us also reflect on our place in the temporal flow: it makes us more aware of the relation between the present and the future. When it comes to consumer objects, some designers have tried to find ways to allow people to make their possessions more personal, in the hope it can foster a connection between person and object that will extend the life of those objects.

One of a range of 'Wasted' beers from UK brewer Northern Monk. The label reads: '2,000 mince pies saved from landfill went into making this beer for you.'

The opposite of disposal, in one sense, is attachment. Martí Guixé, the creator of a range of what might be considered sustainable products, keen to detach himself from an idea of design as the driver of what Reyner Banham called the 'throwaway aesthetic', went as far as describing himself as an 'ex-designer'.[48] What he works on is the idea of attachment. He makes products that can be lifted out of the shorter lifespan that is associated with objects that are valued for their mere functionality, by attempting to stimulate a creative or emotional attachment, encouraging the consumer of such objects to transform them into something that is as personal as it is functional. This, to the extent that it succeeds, can alter the dynamics of waste creation, reducing the desire for replacements and in turn perhaps minimizing the use of resources to make new replacement goods. The problem with the throwaway society was that many consumer products could easily be replaced, which ensured that they existed as finite or temporary possessions within the flux of the inevitable newer and more improved versions that proliferated. In some of the objects and products that he has designed, Guixé deploys a variant on the blank canvas concept, inviting the owner of the object to become its co-creator – the one who ultimately determines what the object is – by adding their own distinguishing marks or touches to the surfaces of these objects. The result tends towards a different durability than might be achieved by making products endure through, for instance, the use of the best industrial materials and product design. This is because Guixé's designs, in terms of their materials, often look cheap. But the idea that a clock or a light can become something more than functional, something almost unique in terms of how the user has contributed to its 'making', opens the way for a shift in perception. One such product, the Do Scratch Lamp, is a black-painted lightbox that the owner must customize (by etching their own words or drawn designs) to allow light to shine through the black surface. This makes it both a functional and quasi-artistic object. Guixé once stated, in reference to being an 'ex-designer', that he did not like the limitations imposed by product and industrial design. He wanted the designed object to transcend its function.

Distractions in the World of Quasi-Objects

There was a famous advert for Apple's iPod, dating from around 2010, which showed an image of hundreds of LP record covers being sucked into the small, portable hand-held device. The idea was that an infinite library could now, to all intents and purposes, be your personal music library. We think of some small electronic devices like smartphones as personal objects, but are they really? The answer, I suggest, is no. They are, rather, relational interface objects; their value is ultimately not determined by what they mean to us or how we treasure them – as people do other objects – but by their usefulness in allowing us to access all that the device is able to serve up through its applications and its ability to access services and content through the interface it provides to a network of infinite proportions. Those services that we access through a device like the smartphone, however, are ever-changing and demanding of more powerful interface devices that are capable of supporting new safety and security features, such as two-factor authentication, bio-identification and so on. So no matter what value we place on the device as a personal object, it is always – as already noted – being made old, being degraded, by what happens on the other side of the relation it establishes between us and a world of seemingly limitless content.

The fact is that smartphones are a perfect illustration of planned obsolescence. They are, as much as any other consumer product, simply garbage-in-waiting, occupying a 'brief interval of currency before their inevitable replacement and transit to the global waste piles of techno-trash'.[49] They are indeed designed to 'break' and fail and ultimately become useless. They are also designed to be easy to dispose of in the sense that a person can quickly – if they take care to do so – erase their identity from the device, rendering it once again an inert functional object. This may seem to enhance the prospect of such a device surviving in a circular exchange, unlike other objects that we have historically filled with personal details (such as diaries, notebooks and so on). But it also makes it more convenient for the owner of the smart device to detach themselves from the object, making it easy to place it into the waste stream (if it is not sold or exchanged or passed on to another user). Even if the smartphone does survive as a 'preloved' or 'used' object in some kind of circular exchange – more likely with rare and unopened early generation examples, which become valued then for reasons other than usefulness or personal attachment – it remains the case that making it easier for these devices to be upgraded or repaired by the user does not solve the basic problem of their relationality. They are, in the terminology of French philosopher Michel Serres, 'quasi-objects' whose usefulness is not contained within the object itself, and whose value is not merely the value we place on it. A 'quasi-object that is a marker of the subject is an astonishing constructor of intersubjectivity', Serres notes.[50] The quasi-object we engage with enlists us, perhaps lays claim to us, our time and attention, pulling us in to complete some task or other or to assist it via those prompts that we must respond to.

And there are other problems with these devices that harm our ability to reduce e-waste: namely that the user who is engaged with their device is probably not given to contemplating the future waste that it creates, given that they are rather caught up in a permanently unfolding present that arrives on demand, through the potentially limitless experiences that are able to be accessed through these interface objects. In short, the user experience is often one that could be characterized as a time out of time, rather than one that is focused on the future. Once again, we might look for solutions to the demands that this media-technology ecology makes on us by thinking of ways to change the nature of our relationship to the world of objects. In 2007 the sociologist Sherry Turkle compiled the book *Evocative Objects*, which focused on certain personal possessions and their owners, and how and why they had become objects that they were so attached to.[51] The idea was, basically, to look at the ways in which our lives are enmeshed in the world of objects. The examples she collected in that book revealed that there is no easy way of separating ourselves from certain material possessions, which at once locate us in 'the middle of things' and connect us to the world in such a way that we feel at one with those objects, just as 'the diabetic feels at one with his glucometer'.[52] The implication when considering an objective such as zero waste is that it may be through emotional kinds of attachment that disposability has to be combatted – objects that perhaps evoke deeper meaning or significance, or indeed are things that people value for more deeply personal reasons.

Such ideas about how objects might be designed in order to extend either their useful life or our attachment to them – aimed at the individual and personal – may be seen to exist on one end of a spectrum that, at the other extreme, extends to the design of larger structures, spaces and human habitats. Design-led innovations today are informed by research into 'the materials and processes of the designed world', which can both reveal new knowledge about resource use and draw clearer links between products, 'the raw materials and natural resources used in their manufacture', in such a way as to acknowledge 'the environmental and human cost of these materials'.[53] Western designers also now look beyond the technology-led model of innovation, which had driven the designed world of the twentieth century, instead looking to 'holistic and sustainable structures and systems' – perhaps ostensibly drawing on some ancient wisdom that sought a more harmonious relationship between the human and the natural, or indeed learning from and copying natural systems 'that have little or no environmental impact'.[54] To take one example in architecture, the Eastgate building in Harare, Zimbabwe, designed by British architects Foster & Partners, aimed to minimize energy waste by mimicking the heating and cooling principles of a termite mound as a means of maintaining a constant temperature and reducing energy use.[55] This kind of biomimicry seems a long way from the more makeshift form of cheapskate architecture described earlier or the austere dwellings that had been the subject of Martin Pawley's *Garbage Housing*.[56] It embodies the ethos of a design movement that has sought to learn from natural systems, while the examples revealed by Pawley almost half a century ago worked towards minimizing energy use through the quite basic method of reusing waste products, taking items like bottles and cans from one sphere and intended use and deploying them as building materials in a different context, thereby taking them out of the waste stream and minimizing the need for new materials at the same time.

As far as living lives mediated by quasi-objects goes, we might point to developments in smart home design, which have been picking up pace since the early twenty-first century and were exemplified by MIT's (Massachusetts Institute of Technology) pioneering *House_n* project (1999): an example of a dwelling that offered 'a curious inversion of the relation between people and their things' – machines become more active, people become more passive – resulting, we might suggest, in a situation where the active human subject might become dispersed in a panoply of smart systems technologies. Thus, artificially intelligent agents become more lifelike (as they take on a series of cognitive and motor tasks) while humans become more thing-like (as they submit their bodies to check-ups, judgements and repairs).[57]

As we emerge into what until recently would have been conceived of as a distant and futuristic existence, such visions of the world to come are as much driven by ideas about how to manufacture environments for sustainable living as they are by the blurring of distinctions between us and our machines in post-human dreams of the future. Thus researchers on the *House_n* project, it was reported, had an awareness that they were not just designing a place to live, but – in a way – redesigning the people who would live in such homes. They were 'exploring ways that a home can use sensors to monitor the home's performance to help people understand the implications of their behaviour on energy consumption'.[58] In terms of working towards a goal of saving energy and eliminating waste (whether that is wasted resources or wasted human effort), such innovations were all arguably extensions of the same logic that produced commonplace technologies such as refrigeration, and even automation evident in sink units for food waste disposal; that is to say, they are extensions of a technological rationality that continually remakes and refines, but in this case also makes more sustainable our relationship to nature.

But offloading tasks onto intelligent systems and devices also runs the risk that we simply become freer to indulge in other distractions that at another level reduce engagement with how we impact the world around us. The expansion of digital memory in smart technologies provides a useful analogy here. The increasing processing power of personal computing devices and the advent of apparently limitless storage capacity that drives revolutions in entertainment media remakes the ways that we may understand leisure time: the logic of limitlessness is undeniable, producing a situation in which a surfeit of possible future experiences – likely energy-intensive ones – is surely the result.

The idea of recycling makes waste management into something uniquely personalized, a new kind of *self*-management. Like medieval ideas about wasted time, it posits the self, the individual, as one who lives in the decay of life. That notion, that waste reduction ultimately has to be concerned with changing consciousness at a subjective level, is wrapped up in the idea of 'zero waste'. There can, of course, be no such thing as zero waste within a human context. Even designs that seek to reduce waste to zero, should they theoretically succeed, still have to give an account of what happens at the end of the life cycle of the designer; of the tools, materials and equipment he or she used to produce such 'zero waste' designs; of the buildings they occupy and the clothes they wear; and of the waste they flush away daily. There is always, in other words, some waste product at the end of any expenditure because, unlike nature and its worms, ants, flies and other parasites, the world we make is abstracted from nature. But, like the promotion of a new idea of recycling from around 1970, what 'zero waste' points to is an attempt to shift consciousness or behaviour through the idea, or the possibility, that there are wastes – or stages of waste – that can be eliminated. As such, 'zero waste' often relies on technology and design and the development of new practical applications of knowledge for the sake of efficiency. These are means to eliminate the unwanted and leftover. Another way of looking at this is to say that ultimately, the aim of such efforts is to conquer time in some way, to disrupt the inevitability of things running down and falling apart.

One reason that a focus on what produces waste and on 'waste consciousness' seems to be desirable is that the phenomenon under observation is then not limited to the easily defined and generally accepted; it is not just defined as the 'wastes' – usually material in form – that we are familiar with, but allows our understanding to be open to its constantly mutating presence, in the sense that waste is then understood a little more as a potential that lurks in everything we do and make, including those things that have not yet been done or made. To be human is to make waste. There is, nonetheless, a tendency to think of waste as something that can be made to disappear, to continue thinking of it as something that is essentially the consequence of errors in ways of thinking and living, rather than as something that is structurally bound to human life and that allows us to locate humanity at the limits of its self-understanding. That is not to say that the generation of waste is not a consequence of human choices that – with hindsight, and notwithstanding the fact that it is not possible to go back in time – we might deal with differently, but rather to argue that configuring waste as a 'problem', or as something that can be solved in some way through the regulation of the behaviours that create their own specific forms and varieties of waste, tends to obscure the fact that if we are not creating one kind of waste, we will be creating some other kind. To be human is not only to create waste, but to recognize when it becomes impossible to ignore and be haunted by it in all its variations, which is something that separates us from creatures in the natural world who do not possess this consciousness. This is because we are at once a part of nature and also stand apart from it. Waste material – particularly that which resists being reabsorbed by nature, such as plastic – is symbolic of this dualism and therefore, as previously noted, occupies a space that represents a realm of non-human human.

The language of sustainability – just like the language of 'recycling' – is quite recent, and likewise dates from the turn of the 1970s, but today informs discourses about our long-term future, linking global politics to everyday habits and practices. But considering that the scope of waste as something that is not only material in form, but a product of a shifting and changing human consciousness, how plausible really are notions like 'zero waste'? Taking the idea to its logical extreme, until we divest ourselves of flushing toilets and systems of sanitation, we will be living outside nature and producing 'waste'. The only way to not create waste is to be part of nature, which is impossible (unless or until one is buried in the earth and decomposing into the constituent elements that nature will then absorb).

It is precisely because of the externality of waste as something lacking value to us that it can seem to negate our existence. It is pushed out of view, allowing us to create a space – an interior life, a disengaged sensibility – where attentiveness to present or future concerns around waste becomes difficult to maintain. No matter how we might define a wasteland, it is not merely an abandoned, depleted or polluted space, but rather manifests itself in those ways because it exists in a particular relationship to us which, in narrowing the perceptual gap between the human and the natural, creates a leftover. Waste, as such, takes up residence in the mobile and the indeterminate, and can often seem to apply to almost everything in transit between accepted categories of value and understanding. But as a spectral phenomenon it is not simply material in nature. It carries us inward, into the depths of recollection and flights of the imagination, where the immaterial, the thinking subject, produces traces that will eventually be recycled for purposes that may remain unknown to us.

Excessive Reality

The idea of the throwaway society was not just the result of a more rapid turnover of material culture, of more goods, products or experiences entering everyday life, but also of a general sense of something having changed; of life having sped up to the extent that it threatened our ability – as individuals, as a society – to maintain control over a world that seemed to be in thrall to a fast-mutating culture of obsolescence.

At the beginning of his book *Future Shock* (1970), Alvin Toffler introduces his readers to the idea that the time they are living in – the end of the 1960s and start of the 1970s – is an unprecedented moment: a point in history when reality has seemingly jumped onto a faster track, to the extent that more events of consequence – so he argues – have taken place in a lifetime such as his own than in all lifetimes before, all the way back to the origins of human history.[59] How such a calculus was put into operation was not really clear, but the essential point he was trying to underline was that everything was going faster wherever one looked, hastening a new and more pervasive sense of redundancy that had spread out far beyond the domain of convenience shopping and consumer experience.

As he introduces readers to some of the themes that he explores in his book, Toffler feels the need to apologize in advance for the likelihood that some of the information in *Future Shock* will inevitably be out of date before long, perhaps even by the time that those very readers are setting their eyes upon it. This, he suggests, was a simple matter of the perishability of fact in a world as shifting and unstable as the one from within which he thought about these things: 'Every seasoned reporter has had the experience of working on a fast-breaking story that changes its shape and meaning even before his words are put down on paper. Today the whole world is a fast-breaking story.'[60]

To choose that example – the circulation of news, the reliability of facts and indeed the status of truth – was an easy way of confronting the reader with the disposability of something they were very much paying attention to at that precise moment. Anyone could see that there was clearly fresh news every day, that if one stepped back to think about it, there was a torrent of information that passed people by. Taken as one kind of information medium, a book – considered in the abstract – is theoretically supposed to be a container for contents that are less disposable than those found in daily newspapers. But the reality that Toffler's book was reflecting on seemed necessarily to be one that would likely continue to exceed the possibility of its capture (a problem even more marked for anyone writing about contemporary culture today), because it was becoming inherently volatile.

More broadly, the 'future shock' was seen by Toffler as something akin to a malady caused by accelerated living. In other words, it was spreading and was no longer confined to the realm of the consumption of disposable objects but had migrated into invisible realms not quite capable of being grasped – into the condition of what today we would describe with the familiar word 'information'. In one sense a vague or indistinct term, absent of specific content, information as data provides a route into considering the unseen complexity of the world we now live in, because once we begin to take notice of it and become conscious of it, the apparently infinite depth of reality starts to become overwhelming. Once upon a time most of us were probably able to ignore the idea that our perception of the familiar world – or some aspect of it, some event that took place in it – might open some hidden passage into a space that was vast and beyond comprehension. This possibility, surely, was one motivation behind the method and approach to the study of nineteenth-century Paris that Walter Benjamin embarked upon in his *Arcades Project* – a work based on the metaphor and materiality of hidden passages and which, appropriately perhaps, was incomplete at the time of his death. And we also see it in reports from other writers who in some way have an affinity with Benjamin and who perform what might be termed acts of literary redemption or restitution in their own way through the orchestration of fragmentary images, ephemera and other apparently worthless details. W. G. Sebald, to name one such writer, would imply that he was driven by a compulsion towards finding more and more bits and pieces to build up a picture of some reality that was able to plumb unknown depths. 'You have a few elements. You build something. You elaborate until you have something that looks like something,' he said. 'And elaboration is, of course, the vice of paranoia . . . it only works through linguistic or imaginative elaboration. Of course, you might well think, as you do this, that you are directing some form of sham reality.'[61]

The contemporary novelist whose work has delved most successfully into such realms is probably Don DeLillo, not only in the previously mentioned *Underworld* – where waste and its many meanings and elements constitute the central theme of the novel – but in *Libra* (1988), his story of Lee Harvey Oswald and the assassination of President John F. Kennedy, and what that event and his research into it revealed about the nature of our desire to rescue every bit of junk, trash and ephemera: that it might allow us to do the impossible and reanimate a moment that is history.[62] Reflecting on the work that writing the novel entailed, DeLillo told the *Paris Review* in 1993 that while the assassination as an event was over in mere seconds, we were 'still collecting evidence and sifting documents and finding people to talk to and working through the trivia'.[63] In a new introduction to the novel, published almost twenty years after its first appearance in 1988, DeLillo noted that new technologies kept emerging, offering the possibility that there might still be some uncovered detail, some hidden fragment of what happened around that particular event, that could finally draw a line under the uncertainties that still swirl around the official account of the assassination. 'Technology by nature, in its brilliant futurity, incorporates a will to surpass the advances of the year, the week, the minute before,' DeLillo wrote:

> Waveform analysis, confocal microscopes, digital replicas. How soon before one technology yields to another? And where, finally, is the truth in this matter? Can some scattered noises in a crowded outdoor setting on a day in 1963 be recovered from an old damaged dictaphone belt, its grooves 75 microns wide, five microns deep? Recovered, copied, deciphered. We want to believe they can.[64]

In our present, digital photography serves as an interesting example of the extent to which technology's potential for capturing the fleeting or evanescent moment produces an excess of objects, digital objects that – even if they are not taking up physical space in any conventional sense, as things cluttering the spaces we live in – still at some point will likely need to be managed or deleted. All the while they of course consume energy, and this creates more unseen waste. Photography in the smartphone age may accord now more than ever with Susan Sontag's notion that it is a medium that ultimately obeys a logic of consumption: one that seeks to appropriate the inexhaustible reality that becomes its subject. 'To consume means to burn, to use up,' she wrote.[65] But it also means to be left in a condition where we will likely come back for more: 'As we make images and consume them, we need still more images; and still more. But images are not a treasure for which the world must be ransacked; they are precisely what is at hand wherever the eye falls.'[66]

The technological apparatus becomes one more thing that reinforces the dualism of human and nature. And that is why the desire to consume what we see through the camera opens up a sense of a reality that is vast beyond comprehension, a reality that, as philosopher Jean Baudrillard might say, becomes hyperreal, that we make more real than real in our attempts to duplicate it. Sontag thought that in the process of engaging with our insatiable drive to consume the world through images, we necessarily end up more detached from reality, 'making it less and less plausible' that we might be willing 'to reflect upon our experience according to the distinction between images and things, between copies and originals'.[67] It was something that DeLillo expressed himself when he told one interviewer that Benjamin's concept of 'aura', and the split it posited between original works of art and their reproductions – which represented a certain loss of aura possessed by the original – was being surpassed by the speed at which technology was attempting to do something akin to duplicating reality. 'There's nothing left but aura,' DeLillo said; 'Reality is being diminished, leaving behind the aura that results from tape-recorders, cameras, microphones and all of the technological equipment that we use to consume reality.'[68]

What will happen to the excess of digital images once the people who have taken them, and perhaps stored them in a cloud-based archive, die? Indeed, if we now live in ways that are largely mediated through digital technologies, then to die is also to die digitally. Methods for keeping the dead 'alive' digitally are already being introduced, with artificial intelligence (AI) bots able to reanimate the departed through the recitation of messages or text documents that they had stored, and words that had been preserved in other computer applications.[69] There is an aspect of data maintenance implied in the creation of digital archives, but even where cloud storage is used as a space to indiscriminately dump one's digital life, or when it is an extension of a user profile on specific digital platforms, there is always an implied need for more storage to maintain the logic of perpetually accumulating data – even if the need is met by digital platforms and services to which a user merely subscribes. There is no stopping the data flow, and most of us who carry smartphones around also now carry, whether we like it or not, the means to generate new data that automatically engages with systems on the other side of the network. The smartphone camera has become one of the main features of these devices. Through it, and the images it generates, we spill out into invisible networks and out into the world, now the source and content of a diversion to be devoured by the eyes of others.

But whatever the content of our digital archives today in the smartphone era, there have always been practical issues to contend with when it comes to data storage – particularly the hidden costs of the time and energy spent in accumulating, storing and retrieving it. For professional archivists, a significant investment of time has to be attached to the costs of managing accumulations of the material objects of this world: data discs, drives and machines to read formats that would eventually become obsolete. The cost in time included the work of transferring legacy files onto new formats (a minimum act to avoid a total loss), and then additionally preserving the machines that would allow files to be accessed in whichever format they finally ended up.[70] There is a strong implication that the storage of digital archives will always be temporary and in need of further labours to safely secure it against future threats of loss as technology moves on. For the archivist, the content of such files may be at risk of degenerating unseen, chewed up invisibly by a process known as 'bit rot'. Bit rot describes a redundancy or uselessness in equipment or software. The document that we see on our computer screen only looks the way it does because the word processing program supplies 'formatting codes that indicate how the words are to look on the page':

> The codes specify typefaces (Times, Caslon, Palatino), type sizes (10 point, 12 point), stylistic variations (italic, boldface, super script), the alignment of text (centered, flush-left, justified) and dozens of other properties . . . When you select the command to italicize a word, the word appears in italics on the display screen, as if the text were simply stored inside the computer in italic type. This is an illusion.[71]

Such elements that allow textual documents to be read as intended – as well as read at all – can become corrupted by the time, age and fragility of the storage medium. But these are not the only problems an archivist may face. So-called 'born-digital' archives (whose content originated in digital forms as opposed to hard, physical ones) are now collected by major libraries. However, the contents of personal or private born-digital archives in particular, by contrast, are difficult to appraise because of 'the disorder in which many records are kept'.[72] This means that the value of the contents can only be assessed following 'a very granular file-level appraisal, which is prohibitively time-consuming'.[73]

There was a time when only writers and artists of various kinds, as well as librarians and archivists and maybe a few select others, had to contend with the problem of a fast-moving reality rendering objects of perceived value obsolete. But on the other side of today's digital and virtual life lie the data wastelands, the latest realm of the cast-off but not yet dead and buried, so to speak. It is familiar enough to us all now that we live through systems, interfaces and interactions that are mediated by invisible networks. These distribute, and amass, a hidden cache of data that, through time, might well end up being degraded in some way, but just as equally – perhaps more likely – remain inscribed somewhere, ready to be reanimated and put to use once again. Information became the content of mediated memory precisely as a consequence of the storage function and potential of the digital technologies that we interact with – this being the primary distinguishing feature of the systems, processes and data archives that our computing technologies allowed us to access, up to and including the first iteration of the World Wide Web. What changes in the era of always-on connectivity, mobile technologies and cloud-based computing is that information takes on another, more evanescent form. From the user's perspective, information is at once the source of an extended (and externalized) 'memory', while consisting of content less durable than that notion suggests: chatter, rumour, conspiracy, the flow of relentless content.

Decades before this era emerged, Marshall McLuhan identified new forms of media with the creation of entirely new environments that we lived within but which became largely imperceptible to us. 'Environments are not passive wrappings,' McLuhan wrote, 'but are, rather, active processes which are invisible. The groundrules, pervasive structure, and over-all patterns of environments elude easy perception.'[74] 'The medium is the *massage*,' he declared in a subtle modification of his earlier and more well-known expression, 'the medium is the message.' What he meant in substituting 'massage' for 'message' was that media constitute environments that, in a manner of speaking, work us over and put us in a relaxed state of receptivity. He explained:

> Ours is a brand-new world of allatonceness. 'Time' has ceased, 'space' has vanished. We now live in a global village . . . a simultaneous happening. We are back in acoustic space. We have begun again to structure the primordial feeling, the tribal emotions from which a few centuries of literacy divorced us.[75]

What McLuhan terms 'acoustic space' is now, in the twenty-first century, that which commands our attention and enjoins us to respond through our network-connected devices. We listen and respond (to the beat of 'tribal drums', McLuhan said elsewhere), and sort ourselves into what have been theorized as 'memetic tribes', rather than think and reflect in a way that would have been characteristic of any previous era that could be said to have been determined in large measure by a technological medium.[76]

By the turn of the twenty-first century the English artist Stuart Brisley, who had spent many years exploring the nature of waste and our reactions to it, had also moved into the realm of the digital and apparently immaterial, producing a work with Adrian Ward titled *Ordure::real-time* (2001). As if embodying McLuhan's comment that it took the likes of artists to create the kinds of 'countersituations' that make visible or 'provide means of direct attention and enable us to see and understand more clearly' how media environments work, they produced an object that demonstrated the unseen processes of decay that can afflict our machines.[77] A large-screen digital projection showed an image of a 'wasting' process as it happened, as blocks of colour shifted and morphed, pixel by pixel, revealing the image as it underwent a process of data corruption that took place in the real time of its viewing (if switched on and off, it started over again). Alongside this screen was to be found a second display, showing the same image in its original uncorrupted state. On both displays the surface and boundaries remained intact, hinting at the fact that whatever processes of corruption or decay do occur in digital media, they do so without breaking down the material containers that convey the content to us. But, in presenting a visual means of observing this process at work, Brisley and Ward show us what otherwise would remain unseen as the machine undergoes internal changes, thereby demonstrating 'the dialectical play between two interconnected states of order and disorder, between generation and corruption, suggesting the potential for change built into any system'.[78]

In a semi-autobiographical 2003 book, *Beyond Reason: Ordure*, which almost shares its title with this digital work, Brisley presents something akin to a temporally compressed journey through a lifetime defined by themes and obsessions that drove his work in the direction of what one can only term 'waste'.[79] Having spent years surrounding himself with the stuff of waste, or confronting human limits in the face of the disgusting, he saw the computerized world as just another facet of reality that, if left to its own workings, tended towards a kind of failure, if not indeed reaching a state of ultimate uselessness. 'Bit rot as bit decay,' he stated epigrammatically, 'unused computer programmes often stop working after sufficient time has passed.' But as his recent digital work had shown, computers may continue to produce garbled 'representations' of data that was once well ordered or able to convey some kind of meaning – which is to say, when in the form and order of its original inscription – but, given enough time, a text file will inevitably decay, almost as if it had been taken over by an intelligence with Dadaist intentions: 'Fo ytitccoapntac sa ansnd oithtce nucof loxen, Ipwhmoicc h owsht ouesedehnot'.[80]

That kind of process, in itself, as we will see, could be a metaphor that allows us to understand how a data wasteland expands out of view of those whose information it consumes – in part because the infrastructure it represents is always the container for systems that are trying to impose some order on what must always be in essence essentially tending towards the opposite, a state of incipient disorder – and thus it grows as a domain that for the most part eludes our full attention.

Brisley was not alone in taking his obsessions with waste into this new field, but what he seemed to be doing was recognizing the essential truth that nothing that is humanly created is able to elude processes of decay and corruption. We might view such work as an extension of a similar trend in other media, where new technology seemed to permit or even demand new ways of thinking about how the world might be sifted and filtered. For French film director Jean-Luc Godard, scraps of old films offered a storehouse of moving images that it was now possible to orchestrate into a poetic reverie to illuminate the apparent proclivity for death and destruction that lay at the heart of human cultures. We see this in the mesmerizing opening sequence of his 2004 film *Notre musique* (Our Music), which – beyond its use of what looks like leftover or degraded film – presents us with another example of waste aesthetics in which the form of the work matches its subject: as a sequence of images of war flash across the screen, scenes of places and people being blown apart and life being reduced to a Hobbesian war of all against all, the quality of the film itself seems to degrade, as if it has been infected with some contaminant that has escaped from the events being portrayed. One might also point to other examples, such as Bill Morrison's film *Decasia: The State of Decay* (2002), which presents an aesthetics of disintegration as a source of poetic reverie for the viewer. At around the same time, other figures became interested in such phenomena as 'glitch' music and 'cracked media', which produced the sound of malfunctioning machinery or equipment that disrupted 'normal' functioning. 'Cracked media', ran one definition, were

> the tools of media playback expanded beyond their original function as a simple playback device for prerecorded sound or image. 'The crack' is a point of rupture or a place of chance occurrence, where unique events take place that are ripe for exploitation toward new creative possibilities.[81]

In one sense, something like 'glitch' music was a way of reclaiming 'waste sounds' for other creative purposes, but what can we say of the reasons why someone would be interested in chasing such elusive noises? Taken together, these were developments that arguably tapped into the same fascination that had arisen in the 1990s with surveillance, which spawned its own particular aesthetic as seen in reality TV shows and police traffic videos, and other evidence of an excess reality that had formerly eluded our grasp. Such things played on paranoid fears about how technology was gobbling up fragments of our lives for some other purpose, but also revealed that, in a media-saturated culture, 'we can only get a charge off "reality" or images which ostentatiously signal reality via their low-tech crudity.'[82]

5

Projections: Wastelands, Real and Imagined

An image of the urban landscape as wasteland belongs not only to the reality of the built environment and its changing forms and structures as it cycles through time and history, but to the human imagination, source of a host of visions and ideas about waste and wastelands. In that respect, it is interesting to consider how ideas about an accelerated present and future are so often reliant on the iconographic potential of waste. In fact, the idea of waste itself becomes the material used in the creation of something else, something new that is taken in and absorbed as part of the culture more broadly.

Images of Conspicuous Destruction

Projections of waste can be identified in particular with twentieth-century science-fiction literature, but one could equally point to its appearance in earlier dystopian or apocalyptic fictions that invent future wastelands. Mary Shelley's *The Last Man* (1826), for instance, envisions the twenty-first century as a time of natural disasters in which humanity is cut down by a plague that devastates the world, leaving it a landscape of ruins, with the eponymous Last Man sailing the shores of the deserted planet with volumes of Homer and Shakespeare his only companions.[1]

When H. G. Wells's time traveller advances at great speed through millennia in *The Time Machine* (1895), he remains in the same location but is thrust into a series of futures that reveal the fate of the human world that he had known. When he eventually finds himself in the year AD 802,701 it is to discover a 'humanity upon the wane', with the structural remains of previous eras unknown to him resting among landscapes that appear collectively as a benign but untended garden.[2] The effect was not unlike what eighteenth-century French philosopher Denis Diderot argued was the essence of the poetics of ruin that he associated with Romantic painting, which is to say, a reminder of how insignificant the human is against the destructive efforts of time.[3]

Later, in another novel, *The War of the Worlds* (1898), a tale set in its own time, Wells shows us the unfolding catastrophe that follows a Martian invasion of Earth. Destruction ensues as the alien presences, taking the form of monstrous tripods, sweep through the human population like a broom across a dust-strewn floor, 'striding over the young pine trees, and smashing them aside . . . a walking engine of glittering metal, striding now across the heather'.[4] As John Carey has written, this would be one of many examples of Wells's fictions to project the world into a wasteland condition, with parts of London and its expanding suburbs being reduced to rubble. Carey notes:

> The destruction is less a matter of human casualties than of postal districts being cleared . . . Towards the end, the narrator walks through suburban London – Mortlake, Putney, Roehampton, Fulham, Ealing, Kilburn, South Kensington – and finds it quite empty of people. Vegetation is returning. A red weed, introduced from Mars, spreads everywhere, burying the remnants of houses in its rampant growth.[5]

For Carey, Wells's visions were uncomfortably in keeping with how the bourgeois literary establishment at the turn of the twentieth century had viewed the masses, with various degrees of revulsion, as almost subhuman.[6]

Perhaps inspired by the examples that Wells had developed, images and descriptions of waste and ruin proliferated in sci-fi visions of the future throughout the twentieth century, an epoch in which the industrialized world seemed to be constantly destroying and remaking its own landscapes, either through war, urban planning or the reinventions of industrial capitalism. By the end of that century much of the industrialized West was in terminal decline, producing vast abandoned landscapes that had been exhausted after yielding up whatever useful or productive potential they once held. In places, what remained, when not transformed into heritage landscapes, were new ruins.[7]

A contemporary preoccupation with the preservation of visual traces of this post-industrial world can be found in a plethora of coffee-table photo books with titles like *American Ruins*, *Derelict Britain*, *Abandoned Places* and so on. There is, in other words, a very particular and abiding interest in ruin and decay, which now more than at any other time can be consumed in the form of images. In the Internet era, for those who may have no personal experience of growing up surrounded by such landscapes, images of post-industrial ruins nevertheless proliferate, even spawning the term 'ruin porn'. Certain places, such as the partly abandoned U.S. city of Detroit – a former hub of car manufacture – have become spectacular exemplars of this trend, inspiring not only photographers but other explorers who, through YouTube DIY videography, document their attempts to break into and explore the condemned or abandoned structures that litter the once thriving 'motor city'.[8] And if the spectacle of these ruined landscapes and cities provides the means to reintroduce us to a world that had been left to rot, it is in cinema that glimpses of waste and pollution

The ruined interior of former automobile coachbuilder Fisher Body in Detroit, Michigan, which has lain empty since 1993.

have supplied the necessary ambience of uncanniness that seems to capture our essential relation to this other possible world that takes form as an imperium of waste. The polluted environment never looked so aesthetically exquisite as it did in Italian film-maker Michelangelo Antonioni's *Red Desert* (1964) – where, in truth, the wastelands and effluvia of industrial production provide the most compelling evidence of the usually hidden environment within which modern life unfolds – or as bizarre as in American director Wes Anderson's film *Isle of Dogs* (2018), in which disease-ridden dogs that threaten to overrun Japan are deposited on a giant garbage island.

While the formula of Wells's *Time Machine* might be said to underpin an entire category of sci-fi writing, it is in the Hollywood blockbuster movie that waste is most spectacularly envisaged, and quite often by choosing settings a short leap ahead in terms of years, so providing scenarios in which familiar places that have been reduced to wastelands can enliven the imagination of the viewer. Sometimes, the switch to the future state of ruin can be achieved without any need to really change the scenery. In the post-apocalyptic thriller *The Omega Man* (1971), Charlton Heston plays Colonel Robert Neville, the last person left alive (or so he believes) following the outbreak of a plague created by biological warfare that decimates humanity, who passes his time in the afternoons by going to an abandoned cinema and screening *Woodstock*, a film he has seen so many times that he is able to repeat the dialogue of the hippies being interviewed on screen.[9] At night he fights off zombies. Otherwise, the city has begun to fall into ruin simply because the absence of people who might otherwise help maintain the streets and buildings has accelerated the sense of decline. Elsewhere, Thom Andersen's 2003 film essay *Los Angeles Plays Itself* – suitably enough a cleverly constructed amalgam of film fragments made to yield up a Benjaminian illumination: that fictional movies carry documentary revelations – contains an impressive section about the glee with which Hollywood has given rise to a 'cinema of conspicuous destruction' whose targets are often its own immediate setting, which is to say, Los Angeles city and county.[10]

One obvious symbolic target for the various terrorists and alien invaders who torment ordinary people in these films is the iconic Hollywood sign itself, which sits invitingly on the lip of the hills that frame the area's picturesque canyons and below whose steep roads run down into West Hollywood, making the iconic landmark visible from Sunset Boulevard below. While the movie destruction of this icon could be seen as an ironic swipe at the tendency of Hollywood films to glorify spectacles of destruction, the sign had in fact already begun to crumble through neglect in the early 1970s, a time of economic crises in the United States. From 1974 until the end of the decade it had been left to languish in its sadly reduced state, a symbol of those 'strange feverish years' in the USA 'punctuated by disaster' and marked by oil embargoes, the Vietnam War and presidential disgrace, until a campaign that engaged local residents and celebrities raised the cash to repair it.[11] At the time, shock rocker Alice Cooper, whose home from home in those days was the bar above the Roxy nightclub on Sunset Strip, where he spent the nights with fellow louche barflies from the rock world, including Keith Moon and Harry Nilsson, gifted the campaign $28,000 to save the letter 'O'.[12]

In his 1990 book *City of Quartz*, Mike Davis, one of the pre-eminent chroniclers of the recent history of the city, detailed the development of *noir*, the literary sensibility that came to be associated with Los Angeles. It manifested, he thought, 'a fantastic convergence of American "tough-guy" realism, Weimar expressionism, and existentialized Marxism'.[13] Looking at examples of this new kind of writing, which he described as belonging to a succession of 'through-the-glass-darkly novels', it was apparent that at some point in the late 1930s such literary works began to become fixed upon the idea that Los Angeles, which had been endlessly promoted as a paradise, was in fact a 'deracinated urban hell'.[14]

When Davis followed up with a new tome, entitled *Ecology of Fear: Los Angeles and the Imagination of Disaster* (1998), he enlarged this vision of Los Angeles as a society (not merely a city) on the verge of seemingly permanent collapse. One way he did this was by looking more closely at its treatment in literary and cinematic sources by undertaking an exercise to examine and enumerate the ways in which the city's numerous imagineers had taken an almost fascistic glee in turning it into a wasteland.[15] Davis claimed that the destruction of Los Angeles in one way or another had been, until the late 1990s, 'a central theme or image in at least 138 novels and films since 1909'.[16] This meant that between that starting point, 1909, and 1996, when he was researching *Ecology of Fear*, 'the city and its suburbs have been destroyed on average three times per year.'[17] As in Andersen's *Los Angeles Plays Itself*, the city of fictional representation has a knock-on effect on how the real city is perceived. There is, unsurprisingly, a widespread view that Los Angeles exists at a point of almost total social and environmental collapse, its dispersed urban sprawl overrun with crime and vice, and harried by destructive fires fanned by the powerful Santa Ana winds: 'post-apocalyptic Los Angeles, overrun by terminators, androids, and gangs, has become as much a cliché as Marlowe's mean streets or Gidget's beach party. The decay of the city's old glamor has been inverted by the entertainment industry into a new glamor of decay.'[18]

In this respect, such a view of the city forms a continuity with the recent fascination for images of modern 'ruinscapes'. It has become, whatever it is in reality, a wasteland of the imagination. Perhaps the most notable of all films in this tradition, for the cult status it was to develop in the years following its release, was Ridley Scott's *Blade Runner* (1982), which presented a vision akin to some latter-day hellscape that Hieronymus Bosch might have had trouble topping, and through which a near-future Los Angeles was able to take root in the minds of millions. That sense of Los Angeles as a place that Bosch might have imagined, had he lived today, was one that struck Jean Baudrillard in 1986 as he recounted the experience of flying over the city at night. In its vast horizontal sprawl, which extends urban grids into the Mojave Desert, the parallel lines and vanishing points of its freeways stretch far into the distance. When illuminated at night, Baudrillard wrote, it presented the city as some kind of 'Galactic metropolis'.[19] The fact that the city's film reincarnation in *Blade Runner* is seen mostly at night, too, only intensifies our appreciation of the visionary power of cinema. Guided by director Ridley Scott's desire to conjure an environment that a viewer might believe represents the future, the 'real' city of Los Angeles is disguised beneath layers of surface neon junk and consumer detritus, which seem to surround and confine our hero Rick Deckard (Harrison Ford) as he proceeds through warrens where market traders have taken up residence beneath the gigantic towering forms of the congested cityscape, dominated by distant ziggurats that house the Tyrell Corporation, which produces replicants to send to off-world colonies.

Scott, perhaps unfamiliar with a century's worth of dystopian visions, thought that since the future that films served up was 'usually presented as pristine, austere, dull, textureless and colourless' he would take things as far in the other direction as possible and disguised his set – which utilized 'buildings' that had appeared in 'half the pictures Cagney and Bogart made for Warners' – with sufficient junk and grime to make what might otherwise have been familiar now unrecognizable.[20] In this future 2019 – a date 37 years ahead of the film's release – the imagery of waste would be all-consuming, as one report from the film set revealed: 'filthy pollution-control devices will sprout on every structure. Enormous dustcarts the colour of sick will stalk the streets spewing vile steam. Cars will be ugly and cramped. A lucky few motorists will be able to own "spinners" – vehicles that can take off anywhere.'[21]

If Philip K. Dick's novel *Do Androids Dream of Electric Sheep?* (1968), on which the film was based, lacked the ability to visualize the cyberpunk dystopia of Scott's adaptation, what it presented was nonetheless also and unambiguously a future wasteland. 'The chairs, the carpet, the tables – all had rotted away; they sagged in mutual ruin, victims of the despotic force of time,' the novel's narrator informs us of the apartment building where the character Isidore lives: 'No one had lived in this apartment for years; the ruin had become almost complete.'[22]

Soon after the film begins, we see Deckard sitting in a crowded market, waiting for a seat to become available at a noodle bar. It's so crowded, indeed, that the cars of the lucky must navigate the city by air. Above, a blimp floats past displaying a large

Rick Deckard, portrayed by Harrison Ford, looks upon a street scene of neon and junk, with thoughts of life in a less cluttered off-world colony advertised by a passing craft: *Blade Runner* (1982, dir. Ridley Scott).

colourful screen with words and images that advertise the benefits of migrating to an off-world colony in order to get out of the overpopulated city:

MORE SPACE
ALL NEW
LIVE CLEAN[23]

This, in the film, is Los Angeles 2019. As a vision, it is tempting to compare it with the way that Los Angeles, and Southern

California more generally, had been marketed almost a century earlier when real-life migrants, no longer having to fear the ordeal of passage through the desert on some godforsaken wagon train, had been lured by adverts rhapsodizing about the California that came out of the nineteenth-century gold rush as the 'Cornucopia of the World'. This was a place 'without cyclones or blizzards' (unlike much of the South and Midwest), in the words of one such advert, with a climate that was perfect for health and wealth.

In *Blade Runner* the contrast between the vision of future health and the decay and grime of life in a city that is dying is all too clear. A punk-looking replicant (played by Daryl Hannah) scuttles for cover in the rain and finds shelter beneath the entrance to what looks like an abandoned building, settling down for the night among the rubbish that has accumulated in a corner by gathering up discarded newspapers to keep herself warm. She is startled by the arrival of J. F. Sebastian, a character loosely based on that of Isidore from Dick's original story, which tells us:

> He lived alone in this deteriorating, blind building of a thousand uninhabited apartments, which like all its counterparts, fell, day by day, into greater entropic ruin. Eventually everything within the building would merge, would be faceless and identical, mere puddling-like kipple [decaying matter] piled to the ceiling of each apartment. And, after that, the un-cared-for building itself would settle into shapelessness, buried under the ubiquity of the dust. By then, naturally, he himself would be dead, another interesting event to anticipate as he stood here in his stricken living room alone with the lungless, all-penetrating, masterful world-silence.[24]

As we see represented in *Blade Runner*, the world outside of this building is, in the words of Mike Davis, 'a sinister entrepôt of alien cultures and races'; a place where the inhabitants of this particular imaginary Los Angeles are as likely to take some monstrous, non-human form – replicants – as they are to be something we might conventionally understand as human beings.[25] The replicant played by Hannah has a look clearly inspired by punk, as if to suggest that the aesthetics or look of punk presented an intimation of the future. Other figures we see move in the film look like clones of the real-life Sid Vicious (who died in 1979), with trademark spiked hair and leather jackets. The film's close-up urban scenes take place on what we might call *Blade Runner*'s ground level, which with its gaudy, neon-and-junk aesthetic signals nothing so much as a consumer society that has been fast-forwarded to saturation point. One might imagine it as a neon-lit vision of Mayhew's London that has been transported to the future.

What always went hand in hand with the imaginaries of ruin produced by Hollywood was the idea of California as a place of pure future, or, in the words of Aldous Huxley, a destination whose refrain was always 'tomorrow, tomorrow, tomorrow'.[26] 'California's emergence as an economic and cultural center revitalized the old millenarian notions of the Pacific as the end of westering History (with a capital H),' writes historian Kerwin Lee Klein.[27] California would be framed through numerous visions during the twentieth century: 'Socialist utopia; suburban paradise; the thousand years of Christ's reign on earth; apocalypse by fire, earthquake, atom, or insurrection; the depths of entropy; the dustbin of history; technological millennium; the destination of the death instinct.'[28]

As far as *Blade Runner*'s allusions to punk as an intimation of a messy future went, Los Angeles, of course, had its own burgeoning scene by the last years of the 1970s. It was captured by American film director Penelope Spheeris in the aptly named 1981 documentary *The Decline of Western Civilization* (a reference to Oswald Spengler's early twentieth-century two-volume work of a similar title). She later described her first encounter with what seemed to be a strange presence that had found its home in some of the city's nightclubs, where people looked like 'insane, crazed aliens from another planet' amid graffiti-covered walls, cockroaches and rats, and 'people lying in their own vomit'.[29] Her documentary was akin to an anthropological document – quite unlike *Blade Runner* – but one that signified the bad future that punk seemed to herald for so many, both in the United States and in Britain.

The Last of England

Punk was not just about a style or a way of dressing, or a particular sound. It represented an entire aesthetic sensibility that was bound up with ideas of waste in numerous ways: trashy clothes (sometimes even plastic bin liners), ramshackle musical performances featuring disruptive or violent audience participation, body disfiguration (the safety pin through the cheek or nose) and boredom, a deep malaise that was evident in exhortations to smash things up, to destroy one thing or another. For the young punks there was nothing to do, no outlets for frustration or excess energy, just the fact that in boredom, 'time piles up as so much waste,' giving it a quality quite distinct from wasted time; rather it is time not utilized at all, time 'realized as waste'.[30] In this environment, wrote Dave Thompson, who witnessed punk sprout in 1970s London, all the usual values seemed to have been reversed: 'Bad is good. Garbage is valuable. An anti-world. Just like in the *Superman* comics that I read as a child, where it was called the Bizarro World.'[31]

Thompson was not alone in thinking this strange eruption and celebration of chaos was a form of aberration. The responses that greeted some of the first encounters between the phenomenon of punk and the writers, reporters and others who had some role in gauging shifts in the cultural weather of 1970s Britain served up an image of impending social collapse. The punks were part of a generation for whom 'pornography was just another comic book' version of life, they said, unburdened by the need for social decorum or manners. Their mode of expression was infantile defiance. No one was really sure where they came from, but as far as could be made out, they seemed to attack everything around them without provocation. They even had difficulty being civil to each other – every other person was a 'fucking cunt'. Their figurehead was a singer who couldn't sing, 'a vituperative ball of acne'[32] who possibly named himself in honour of the fetid air that hung over the stagnating seventies: 'Johnny Rotten'.[33] He performed, sometimes, with his head dangled almost upside down, pulling contorted shapes that gave him the appearance of the hunchback of Notre Dame. Rotten was so bent on disruption that he might as well have devised a way to kick himself in the stomach, because self-destruction was going to be his only achievement. Almost as strange was the sight of the creatures who followed behind him. Their number included a great many who, like Rotten, had ditched their proper names, already prepared for the new world to come: Rat Scabies, Sid Vicious, Poly Styrene.

The whiff of disorder was always present when punks appeared onstage to perform their degenerate version of rock'n'roll – it might have been the same old second-hand riffs, but it was important to note that it had been invigorated by being performed in a way that was markedly worse than previous versions. When Rotten and the Sex Pistols were finally able to perform after either being banned or forced to audition for local councillors in various cities across the country, it was often under a hail of spit or missiles, in the face of torrents of abuse. Their followers, who often took pains not to treat these specimens of the near future as stars or heroes, looked and acted themselves like Martians who had recently arrived on our planet. They tended to stand around after performances rather than dispersing and going home – because this wasn't entertainment, it was life. Where was home anyway? These were the people who wanted to destroy society and start again. They sang of, and sounded like, anarchy, apathy and violence all at the same time. In fact, so successful did they become at projecting the end of civilization as the world understood it that, by 1977, just a few years after they first appeared, they had to be silenced and tormented until they went away or killed themselves.

But the spectre of punk as the intimation of something broader and more closely connected to a sense of decline was not just felt at its source, in London, but soon communicated to foreign lands. For New York's *Village Voice* the messianic potential of the by then infamous Sex Pistols, some of whose members were clad in ragged Vivienne Westwood shirts that bore the word 'Destroy' across the front, was clear. 'Johnny Rotten embodies an apocalyptic narcissism,' noted one reviewer; 'his mocking, gloating laughter echoes across the ruins. He's scabrously charismatic: the murderous gleam in the eye is a turn on, and droogy cultists will imitate his every trashing move.'[34]

This emanation from the ruins of post-industrial Britain was felt as far away as the Soviet Union, where the phenomenon attracted the attention of the Communist Party, who clearly thought that the reverberations from the fevered coverage of the British tabloid press might be heard in some way by their own, possibly alienated and impressionable, youth. 'The music and the lyrics of punk rock provoke among the young fits of aimless rage, vandalism, and the urge to destroy everything they get into their hands,' wrote *Moscovsky Komsomolets*, the paper of the Young Communist League, decrying the whole thing as essentially an anti-communist ruse.[35] For music critic Greil Marcus, in his epic genealogy of punk, *Lipstick Traces* (1989), what was going on could be assimilated within a shadowy network of artistic and political forces, heretical and iconoclastic, and with destruction at its heart, reaching far back into the past.

Marcus saw the most recent antecedent of the doomed Sex Pistols, who sang of a future foreclosed, referring as much to themselves as to the prospects of anyone who was listening, in the Situationist International. Situationists like Guy Debord and Ivan Chtcheglov warned the architects of the new urban world that had risen up in the aftermath of the Second World War that from these landscapes of grey concrete – whose houses, apartments and underpasses were designed to house, contain and channel the movement of residents who as a result became akin to subjects of an experiment – would 'come the last troglodytes of the slums and the ghettos'.[36] In the British context, and particularly in London, punk was also like something out of Mayhew or Dickens projected a century into the future.

One route out of the filthy city of the nineteenth and early twentieth centuries was found in the growth of the modernist urban developments of the mid-twentieth century. The shift that took place in the nature of place, leaving behind the familiar streets and haunts of Victorian towns and cities in favour of the blankness of a concrete world of looming towers and elevated streets, also allows us to understand how a new era of cleanliness and, thus, a new era of waste was birthed. 'Hot water by Sadia means no dirt, no work, negligible maintenance, and utter reliability – all at lowest running cost,' declared a 1961 advert for Sadia Ltd (suppliers to British local authorities and their new modernist estates) showing two well-dressed male town planners surrounded by happy residents.[37] The movement from slum to Modernist estate also charts the changing nature of wastes – we move from filth, dust and putrescent matter to rubbish and garbage. The tower block allowed for greater concentrations of people in a place, resulting in more concentrated volumes of waste, but also the development of new means of automatic disposal.

As Owen Hatherley observes in 2010's *A Guide to the New Ruins of Great Britain*, it took a while for a reaction to these new living environments to emerge, but when it did manifest itself, 'the cities of tower blocks and motorways' became associated with punk, 'which defined itself as having come from those towers and walkways', even if this was mere projection – an attempt to mythologize punk's origins.[38] The stupefaction of modern life, according to the Situationist Chtcheglov, was able to take hold because life had become boring and banal, and people were now obsessed with consumer goods and the labour-saving products of modern living. 'Everyone is hypnotized by production and conveniences,' he declared, 'by elevators, bathrooms, washing machines . . . Presented with the alternative of love or a garbage disposal unit, young people of all countries have chosen the garbage disposal unit.'[39] And indeed, as punk swept across London in 1976, there were those who could content themselves with life in a new concrete high-rise, complete with its kitchen waste disposal system, even if the concrete world was already turning to ruins around them.

During the summer of 1976, when punk began to break into the public consciousness, the UK was broiling, dazzled by the longest heatwave in recent memory. 'Equatorial' was how Martin Amis described it in a review of Anthony Burgess's futuristic novel *1985* (1978), which projected England into a future 'kakotopia' – literally, a wicked place – that seemed to have arrived a decade ahead of schedule and was 'boiling up, partly with crisis, partly with rage'.[40] As J. G. Ballard, himself a visionary of the apocalyptic mode, said of the punks (sounding not unlike Chtcheglov): 'bourgeois society offered them the mortgage, they offered back psychosis.'[41] Self-published fanzines subverted the rules of graphic design and practice to create a style that could be identified through 'rough photocopied images, hand-drawn letters, ransom-note lettering, crudely cut and torn edges' – a trashy aesthetic, in other words, which has since 'become a category in the history of graphic style'.[42] They made, in the words of cyberpunk author William Gibson, 'cut-and-paste analog artifacts'.[43]

The look of punk as something to adorn the body, as a style or fashion, was also bound up with ideas of worthlessness, trashiness and the idea of 'recycling' certain sartorial elements into something new. It was a form of bricolage, noted the cultural theorist Dick Hebdige in 1979 (borrowing the term from the structuralist thinker Claude Lévi-Strauss), but one that departed from the borrowings of earlier British youth cultures inasmuch as, with punk, the objects were taken from 'sordid contexts' and consisted of 'trashy fabrics', 'vulgar designs', 'nasty colours' and 'obsolete kitsch', thereby offering – intentionally or otherwise – 'self-conscious commentaries on the notions of modernity and taste'.[44] Hebdige was writing at a time when it was possible to step back and make such an observation, a couple of years after the general elements of the punk look or style had migrated beyond its ostensible origins in the London music scene of 1976–7. But it should also be noted that prior to the first appearance of these elements of style in the UK, a similar 'ripped and torn' look was worn by Richard Hell of Television (and later the Voidoids) in New York in 1974. Hell would claim that Malcolm McLaren and Vivienne Westwood took his look and complemented it with their own futuristic-yet-backward-looking creations, such as trousers with bondage elements – straps and buckles, as if taken from a straitjacket – that could make them useless for getting around in. Others, like Johnny Rotten, spontaneously developed their own look of chaotically chopped hair and selectively torn or defaced clothing items, making the punk look – certainly by the time of 1982's *Blade Runner*, where punk figures are part of the future landscape – a quick and easy way of signalling an impending bad future. While this visual identity had come from the music scene, it is also true to say that earlier uses of the term 'punk' referred to states of abjection or worthlessness: junkies, male prostitutes and the insalubrious and obnoxious residents of the urban street.[45]

The ground zero of the punk look was 430 King's Road, London, site of the boutique fashioned by McLaren and Westwood, and the place where punk rock – in the form of the Sex Pistols, the first of what would soon become dozens, then hundreds, of bands – took shape much as in the prophecy suggested by Greil Marcus in *Lipstick Traces*. During one of its frequent reinventions, this boutique changed its name from 'SEX' to 'Seditionaries' (the latter clearly derived from the term 'sedition' but also likely intended to allude to the words 'seduction' and 'revolutionary'), shifting the emphasis of the guiding idea behind McLaren and Westwood's activities into slightly more political terrain.[46] Gone were the giant pink letters spelling SEX that had previously occupied the signage space above the entrance, as the shop's appearance was replaced by something stealthier and more intimidating looking: its front windows and door were now caged in wire grilles, as if it was the entrance to a secure unit or a place where the authorities locked up criminals and people that presented a risk to the public. The interior of the revamped space was dominated by two huge wall-sized images that pictured transformed cityscapes. One was a photograph of Piccadilly (one of London's main tourist draws) positioned upside down while the other depicted a city destroyed, an image of Dresden, Germany, in the aftermath of the firebombing raids of 1945.[47] On the clothing racks the labels read, 'For soldiers, prostitutes, dykes and punks' – a litany of those perceived to be disposable, one might say – which was an imaginative projection of the possible advocates of seduction and revolution: the ideal subjects who might sport the confrontational, ripped-and-torn aesthetic that came to visually define the punk era in Britain.[48] For the mainstream, such people could only be regarded as the 'inadequates' of contemporary youth, reported one newspaper, those 'born from unemployment and poverty, dressed in ugly, throw-out clothes, fostering a hazy belief in the destruction of all things establishment'.[49]

Jon Savage, author of the history of punk in that era, *England's Dreaming* (1991), insists that in the decades that followed, everyone misunderstood the consciousness of those times. The point of the shabby ripped-and-torn aesthetic, he argued, was that it expressed a *futuristic* attitude at the heart of punk, whose adherents embraced sci-fi visions of future ruin as the landscape and playground on which they might reinvent who they were, if not the idea of the human being.[50] Thus, Seditionaries, Savage argued, was 'a futuristic self-contained environment that put together all of McLaren and Westwood's obsessions – Sex, Royalty, England's atrocities – in an extreme environment'.[51] This idea of punk as a super-futuristic phenomenon was equally evident in its cannibalizing of other eras and numerous, often disparate, styles for its 'new' forms of bodily expression. The punks of mid-1970s London, before the phenomenon spawned copycat looks that were sold in high street stores, and befitting a culture in which the flea market became an important location, 'sourced their look' from various antecedents ranging from science-fiction visions to 1930s Weimar Germany, pornography, the films of Andy Warhol and 'bondage trousers that forced you to mince and hobble, skirts made of black PVC, diamanté, plastic macs, winkle-picker boots'.[52] As can be seen in images from their earliest photoshoots of late 1975, the Sex Pistols themselves wore new fashion items: clothes by McLaren and Westwood that looked unlike anything that had been seen before.

These looks, of course, were also sported by their designers. A 1977 profile of the newly famous Westwood observed reactions to the way she looked and dressed on the street and at work in the boutique, a space in which she seemed securely at home. By contrast, on the streets of London's Kensington, a sometimes incredulous world looked on. 'Her bleached hair is styled upright like a soapy shaving brush, her makeup is shocking pink, smeared around the hairline,' the piece noted, and 'dangling from her ears are small silver penises. As she goes by, a Jaguar and a lorry have a near miss . . . while two staid ladies emerging from a supermarket mutter "disgusting, quite revolting".'[53]

Westwood thought her appearance 'intergalactic', something that made her feel like a 'princess from another planet', evoking the way punk London represented a shift in consciousness and the jarring of temporalities.[54] England's past glories was one undercurrent conveyed in the media coverage of the 1977 Silver Jubilee celebrations for Elizabeth II – but the reality was a London pockmarked with wastelands; a punk London that existed as a space in which the possible could accelerate into the actual, where the counterpart to the Jubilee imaginary was the sound of the Sex Pistols lambasting imperial illusions in the song 'God Save the Queen'.

Westwood's boutique had, during its previous incarnation as SEX, seen her charged under old, arcane legislation – the Vagrancy Act of 1824 – for displaying a T-shirt that showed two male cowboy figures with exposed penises rubbing up against each other. Record shops whose window displays featured the Sex Pistols LP *Never Mind the Bollocks, Here's the Sex Pistols* were also charged with obscenity under the same law, something else that seemed to connect punk to a vanished Victorian London. When this Vagrancy Act was conceived, it had applied to the actions of the kinds of characters who might have been brought to life by Henry Mayhew: 'rogues and vagrants, fortune tellers and tramps, or any person exposing wounds to gather alms, exposing his person, or exposing in any public place any obscene print or picture, or obscene exhibition'.[55] The LP record sleeve consisted of the words of its title in black, made to look like a ransom note, against a Day-Glo yellow and pink background. The word 'bollocks', soon to be the subject of an absurd obscenity trial in which the judge declared he had to reluctantly find the Sex Pistols and their record company not guilty, was the offending element in this instance.[56]

The punks, of course, were all children of those who had lived through the war, parents who believed that the chaos and destruction of those years were behind them, replaced by peace and prosperity. But the signs were not good – the 1970s were promising to be gloomier and more complicated than the '60s, which had already been a period of upheaval. And what followed that decade? Kids with hacked-off hair, deliberately slashed and torn clothes, and disrespectful attitudes to the older generation who had lived through the war – something symbolized by the confrontational appearance of swastikas as punk regalia, which was seen as mocking the older generation's wartime experience. Perhaps the recourse to Nazi symbolism wasn't just designed to shock or to illustrate that the generation gap was real, but to say that the world that had been won in the war was not what they wanted. As Westwood said in a 1977 interview, 'the point about punks is that they change their ideas and behaviour all the time. The one conviction they all share is that they want something other than the dull, safe, trapped life-style they see around them.'[57]

Prime Real Estate

When Russ Meyer, the American director of trashy films like *Beyond the Valley of the Dolls* (1970), scouted London for locations that might be suitable for the Sex Pistols film that Malcolm McLaren had hired him to make, he was particularly inspired by the landscapes around London's docks, which presented a post-industrial vista of dereliction on which stood the huge abandoned Victorian brick warehouses erected when the Thames was booming with global trade. He imagined an opening scene where Sid Vicious, the member of the group who could most easily embody an idea of offensiveness, would emerge from a sewer in Wapping, arriving like some creature from the underworld into the vacant passages of an empty warehouse district.[58] Some of the locations around the same docks had already featured in a 1976 episode of the popular British TV series *Survivors*, where the landscape appeared as a stand-in for a post-apocalyptic world that had come out of the other side of a global plague and whose survivors had retreated underground, into the rat-infested tunnels of London's Tube network.

In terms of what it said about the built environment around the river, that fictional scenario might not have been too remote from the actual facts. Following the first dock closures in 1970, it was not long until 'the banks of the Thames were bare and empty',

> with echoing warehouses and waste ground the only visible remnant of what had once been one of the city's glories . . . the wasteland of the dockside area, once the centre and principle of the city's commerce, was in a larger sense an emblem of London in the 1970s.[59]

The dockside environs may have fallen into disuse, but the abandoned land and spaces around the river were nonetheless seen by some as a suitable location to develop their own ambitions, making use of the structural remains of the commerce and trades that had now vanished. Indeed, the Sex Pistols' film-maker Julien Temple had first encountered the band as they were rehearsing in an abandoned warehouse near Rotherhithe, when he was wandering aimlessly around the docklands landscape one day in the summer of 1975. The following year, the *Daily Telegraph* reported, an enormous, deserted expanse of land around the Surrey Docks at Rotherhithe was being transformed into what journalist Clare Dover described as a hippie trail. Here were 'people wearing shawls and headbands, and scruffy barefoot children playing in the dust'.[60] It was like a scene from a shanty town, with plastic tents scattered around on the land that had been temporarily given over to this new population by the Port of London Authority. 'Some of the wasteland is being turned into allotments,' Dover reported, but it was proving difficult for these new urban farmers, 'a struggle in which nature seems to be winning'.[61] This was part of a week-long festival of alternative living titled 'People's Habitat'. The Surrey Docks site, said a notice advertising the event, stood as 'an example of the destructiveness of our capitalist society'. Organizers hoped that the festival would give people ideas of how to make use of such wasteland, and laid on workshops about urban farming and 'the construction of wind generators, solar panels and other alternative technology equipment'.[62]

One keen observer of what was happening in London in 1976 was another young film-maker, Derek Jarman, who had taken up residence on the south bank of the Thames at a location known as Shad Thames. At Butler's Wharf, in an empty dockland warehouse languishing amid the crumbling remains of a nineteenth-century landscape and close to Tower Bridge, he became part of a community of artists who had started moving into the area at the end of the 1960s, taking over whatever suitable abandoned buildings they could find. For the new owners of these properties, spread across an expansive wasteland that now consisted of seemingly redundant real estate, the trick was to hold on to the land and whatever property remained in good condition until someone came up with a way of exploiting its latent value. They were willing to accept whatever meagre rents they could – usually from artists who sought out these buildings for the large, open interiors that could also serve as studio spaces – while they waited for some sign of a hoped-for redevelopment scheme.

One night in February 1976, by some twist of fate, Jarman found himself at a party that had been organized by another Butler's Wharf resident, the sculptor Andrew Logan, when Malcolm McLaren and Vivienne Westwood arrived with the new 'pop group' they were promoting. Jarman filmed a couple of minutes of a performance by the group, marking the first time the still unknown Sex Pistols had been caught in action, a performance that also featured the futuristic-looking Jordan (real name Pamela Rooke), who ran the McLaren and Westwood boutique on the King's Road. She was to become a figure of interest to Jarman as he set about making his film *Jubilee* (1978), a fable in which punk-era London was projected into the future – not quite the thousands of millennia of Wells's *Time Machine*, but a modest four hundred years – to provide an allegory of social collapse. *Jubilee*'s punks are characterized by the nihilistic violence of a gang of young women, led by Mad (played by Toyah Willcox) and Amyl Nitrate (played by Jordan), who roam the streets to a punk rock soundtrack. They assassinate the queen, 'while her namesake, Elizabeth I, is transported through time to see what has become of the city that became so powerful during her reign'.[63] Part of the significance of punk, Jarman thought, was the fact that it was all happening within the context of the celebrations for the Silver Jubilee. That event, to his mind, epitomized the fact that whatever else the British royal family represented, they were in themselves 'a great symbol of bondage because they are completely trapped', as he told Jon Savage. 'You cannot think of any individuals who are more in bondage to the past.'[64]

The prospect of destruction that punk seemed to promise was, within the larger context of a century of barbarity and destructiveness, relatively insignificant. There was, nonetheless, a spectre of ruin in its headlong rush into obsolescence, and something about time and timing at work in the impact that it had. London in the 1970s, like many towns and cities across the country, could at its worst appear to be a shabby and dirty place. Think of the slum clearances, the squats, the industrial grime still apparent on great buildings that have since been bleached clean, the piled-up rubbish and, in Liverpool, the unburied bodies of the so-called Winter of Discontent (a period between November 1978 and February 1979 marked by widespread strikes, when the sense of national decline in Britain was at its most acute). Also present were the former Second World War bomb sites, which, even if they no longer existed so much in reality, still haunted the imagination through the popularity of the TV series *Danger UXB*, which was set in the early 1940s and followed the exploits of a team of army engineers whose job was to go around defusing the buried bombs left after German bombing raids.

The title of Jarman's later, and similarly apocalyptic, film, *The Last of England* (1987), came from Ford Madox Brown's painting of the same name (1855), which portrays hopeful emigrants setting out by boat for Australia, the white cliffs of Dover in the background. The painting was made at a time when leaving behind one life in the hope of finding something better in the new world was the only option for many who wished to begin over again, and one that many Britons took up eagerly. In 1852, when Brown began work on the painting, emigration had reached a peak.[65]

In Jarman's time, the sense of a London that had been abandoned was not difficult to project if one sought out the right places, particularly those around the Thames and the old docks, but his film also revels in an aesthetic of waste – of the leftover, broken and abandoned.[66] In fact, as well as its use of locations such as the Royal Victoria Dock and Millennium Wharf, *The Last of England* carries that aesthetic in its style and construction, looking as if it could have been put together from a collection of scrap film (and played through malfunctioning equipment) along with the notebook fragments that make up the narration. 'Weep and mourn for the last of England,' a voice intones over melancholy music. The sounds of disorder and incipient violence accompany jittery film of a ruined landscape. 'Sieg Heil', chant an unseen crowd, their voices mixed in with the defeated sound of the Speaker of the House of Commons calling, 'Order! Order!'

against a cacophony of babbling voices, the sound of chaos breaking out everywhere. The figures we see being pushed onto a boat at gunpoint in Jarman's version of the Brown image are caught at a transitional moment. However, in this version the means of escape seem much less certain than in Brown's, even if the last scene – where the sight of the figures crowded onto a small boat could allow us to interpret the title of the film as more of a 'goodbye to all that' than something more negative – offers the possibility that they had found a way out.

While it was a film that made use of derelict sites around the docks, *The Last of England* was both a return and a final goodbye to the docklands, the place he had lived for around a decade beginning in the late 1960s. His tenure at Butler's Wharf had, in fact, been brought to an end by the decision of the developers who owned the land and its former warehouses to move on the artistic community that had established itself there over the previous decade and had done much to keep many of these properties from falling into total disuse. The transition that this entailed serves as yet another instance of how the wasteland becomes fertile ground for something new, if the timing and conditions are right. Jarman would later become known for taking up residence in another wasteland, when he took over a property adjacent to the nuclear plant at Dungeness, Kent, creating a garden around a cottage in what looked like another post-apocalyptic landscape.[67]

One of the most vivid documents of London's docklands in the period can be found in artist William Raban's *Thames Film* (1986).[68] Using film shot during journeys along the Thames, often at dawn and dusk, Raban weaves together facets of the life, death and rebirth of the river to show us the docks as a site of ruin and vanishings. In the mist, the dark shapes of empty piers loom into view as if at the end of the world; half-submerged ships rest, surrounded by large containers that float on the surface of the water like unclaimed flotsam after a battle. Old photographs and etchings are occasionally intercut to appear like flashbacks, the picture zooming in on details that one might easily overlook to reveal tiny figures on the edges of the water, working and playing, before we are returned to the steady progress of Raban's vessel and the view from the river in 1986. What we see is a survey of a dockside composed of rotting timber wharfs, rusting hulks of abandoned vessels, and doom-laden machine sounds that echo across a desolation of rubbish dumps, scrapyards and what seems to be the back end of civilization; the world turned inside out, used up and left for dead. There is an overwhelming sense of the sheer amount of material and work that went into building up the river as a place of commerce; but also, the sense of the river and the docks being returned – even if only temporarily – to nature, and the deadly work of time, which seems to be breaking down or turning to rust anything that has been neglected. Raban uses the etchings from centuries past to show the river as a place of death and execution, the dead left to hang, Raban's narrator tells us, 'until the tide had passed three times'. As if nature were the co-executor of maritime justice.

A period of clearing away and beginning again – part of the decades-long regeneration that marked the era – was by then under way. 'London's wasteland', ran one newspaper headline about the abandoned docklands in 1981, offered '5,000 acres of new hope'.[69] The newly created London Docklands Corporation was moving in. The 1980s, in fact, would be a time of redevelopment across many desolate former industrial landscapes of the UK. The construction firm Lovell splashed its offer of urban renewal in partnership with local authorities to the owners of those long empty ex-industrial landscapes across a full-page newspaper advert, accompanied by a vision of an empty site strewn with rubble that might have been procured from one of the country's leading landscape photographers:

In 1986 Margaret Thatcher fronted government-backed regeneration initiatives to revive the landscapes of industrial dereliction that by then could be found all over Britain.

> If the sight of a private developer working to revitalise an urban wasteland means only one thing to you, then stand by. The capitalist plot thickens! For almost 20 years, a certain private party has actually been working alongside scores of local authorities bringing much needed regeneration to docklands, industrial wastelands, greenfield sites and inner city areas.[70]

It would not be long before the British prime minister, Margaret Thatcher, as midwife of the new regeneration schemes, was seen undertaking photo opportunities in the corresponding wastelands of the North. She was seen walking across scarred and ruined landscapes around Teesside as more development corporations and enterprise zones were launched in numerous former industrial heartlands across the country, the forerunners of other nascent heritage developments on lands formerly given over to mining and shipping: relaunched as cities of culture and sites for garden festivals.[71]

The Trashman

It is instructive to inspect Ordnance Survey maps from the third quarter of the twentieth century to understand just how close London still seemed to be to the wreckage left by the war. If one looks closely, it is possible to see areas on the maps pockmarked with the label 'ruin'. Some of these sites of dereliction had been there since the end of the war, of course, while others identified the locations of buildings and plots of land that had been abandoned and left to decay for one reason or another. The German-born artist Frank Auerbach, who had arrived in England as a boy and witnessed 'building sites multiplying across the bomb-scarred city', studied art at various institutions in the 1940s and '50s, becoming obsessed with London's bomb sites.[72] After he took up residence in Camden Town, he 'spent years sketching the workers clearing dangerous sites' around the city.[73] Auerbach would be one of the sources for the character Max Ferber in W. G. Sebald's novel *The Emigrants* (published in 1992 in German, 1996 in English) and a fitting conduit for Sebald's meditations on how all life in the twentieth century could easily have seemed to be unified in destruction.

By the early 1950s, when Auerbach was still a young artist, the ruins left by the bombing raids of the 1940–41 Blitz presented the city as something analogous to the Romantic sublime: 'a marvellous landscape with precipice and mountain and crags, full of drama'.[74] Fifty years after he first made these drawings and paintings, they were brought together in an exhibition at the Courtauld Institute in London in 2009 under the title 'London Building Sites, 1952–1962'.[75]

It was amid one such landscape around London's Camden Town that the artist Stuart Brisley found himself at the start of the 1980s, following a period of more than ten years during which he had become known for his physically demanding and often repulsive performance works. His work shared in that era's rejection of the museum and gallery as the legitimating spaces of art in favour of events that were alive and characterized by temporal duration, taking spectators 'into conditions of immediacy where attention is heightened, the sensory relation charged, and the workings of thought agitated'.[76] According to one early textbook looking at the development of art environments, happenings and performance, Brisley was creating works that were more akin to dark and 'often violent rituals'.[77] Many of these manifested a very particular and persistent attempt to get in touch with the abject depths of human experience, to push what it meant to be human right up against that which we would designate as the non-human human through the exploration of physical and psychological limits. Although Brisley would become known as one of the leading exponents of this type of performance art, it was, he later thought, a time when his work was caught up in the idea of disgust as much as it was concerned with performance itself.[78]

But although moving away from the extreme physical challenge of live performance, Brisley was nonetheless still drawn to subjects that pulled him to the margins of what was artistically acceptable. This led him naturally to the zone of dereliction that he found on his doorstep. To the east and south of Camden's Georgiana Street – where he began living in the late 1970s – was a sprawling topography of railway yards, brownfield sites and bombed-out spaces left over from the war, as well as a skyline that was dominated by the iron frames of a number of Victorian gas holders, by then mostly the relics of an older system of energy distribution – whereby gas was extracted from coal and stored in these structures – that had been superseded by the delivery of North Sea gas to Britain through pipelines. These gas holders, relics of an era fast disappearing, remained standing as iconic symbols of a dwindling industrial landscape, landmarks in the wasteland between central London and Camden Town. An architectural appraisal of the area, written in 2007, just after the plans for the new Eurostar station that occupies St Pancras today had promised to finally kick-start the redevelopment work, noted that the space behind King's Cross – with its disused marshalling yards and pockets of ruin and rubble-strewn ground that seemed to have been abandoned since the end of the war – stood as one of the worst examples of how difficult it had proved to bring life back to parts of London that had fallen into this state. 'For decades', it stated, 'attempts have been made to produce schemes to urbanise [an] area that has been full of seedy uses, totally inappropriate for the middle of a great city.'[79]

Closer to Brisley's home were the remains of a large power-generating station formerly operated by St Pancras Borough Council, which was previously fed with the output of a large waste destructor at the same location. The heat from incineration created steam that was used in electricity generation. Now

King's Cross, London, early 1980s. Gas holders sit behind a bulging scrapyard.

partially transformed for light industrial use, it stood on the west side of Royal College Street. In the near distance, beyond the gas towers, was King's Cross Station itself, situated in an area that was a notorious haven of vice, lowlife street culture and murder (the area, with gas towers featured prominently, provided a suitable backdrop to the 1955 black comedy *The Ladykillers*).[80]

In the 1830s, when land purchases to clear the way for the railway were made, the area around and to the north of King's Cross and St Pancras, while originally intended to provide 'suburban dwellings for the industrious artisan', according to John R. Kellett's 1969 book *The Impact of Railways on Victorian Cities*, had 'declined even from this modest intention into an area of fourth-rate residences, public and boarding houses'.[81] The station at King's Cross – opened in 1852 – was still under construction when the first volume of Henry Mayhew's *London Labour and the London Poor* was being published, but the characteristic seediness of the area – which lasted into the twenty-first century, helped along by the presence of the new railway station, a place of transit and shifting populations – was, of course, not overlooked in his investigations. It was a place popular, he noted, with pickpockets, who were to be found 'dangling about' near 'beershops' with other members of the criminal fraternity.[82]

Beneath the streets of Camden Town, to the north, flowed a tributary of the Fleet river, described by Mayhew as 'the most remarkable sewer of the world', and which by the time he was surveying London ran underground through much of the city, from its origins north of Camden in Hampstead to Blackfriars Bridge, where it emptied into the Thames.[83] At Camden Town, the Fleet carried away the sewage of countless streets before emerging into the light again and, Mayhew noted, 'the open ditches of Kentish-town, Highgate, and Hampstead'.[84]

By the turn of the 1980s, when Brisley was living in Camden, the area was roughly halfway between the end of the war and the regeneration of the area that culminated in the opening of the St Pancras Eurostar link in 2007. As well as the confluence of these features of this urban landscape, which gave it the air of a place unlikely to be on the itinerary of most visitors to London, Brisley found that he had also arrived at a location that was situated between a number of homeless refuges. For a long time he simply observed the movements of the vagrants from the window of his home. But later, as he was feeling his way around the neighbourhood, he would discover that the only real community he came into contact with in the area was the one made up of those who moved between the various hostels and who otherwise occupied the abandoned zones that constituted much of the landscape at the time, either sleeping rough in the neighbourhood or using it as living space during the day. One thing that caught his eye was that what he thought of as 'indoor living' seemed to have found a place here, with shabby furniture appearing in those in-between spaces that most people tried to avoid: 'In the summers people would gather settees, chairs and beds, which was the rubbish along the canal and from nearby, to make temporary living spaces.'[85]

And so, just as he began for the first time in more than a decade to think of moving away from live performance, Brisley started recording and collecting evidence of what he saw around him every day, first with a camera and later by gathering up odds and ends that had been dumped by other members of the public. Some of the images from that time show piles of rubbish in the form of discarded personal objects, as well as immobile bodies, passed out or asleep – the first of the evidence that made him aware that he was living in the middle of what had become a de facto dumping ground not just for random rubbish but for the destitute.[86] It was a place where unwanted things and displaced people were to be found, the descendants in some way of Mayhew's itinerant wandering poor, here more than a century later as evidence that poverty – like waste – never really disappeared.

As an artist, Brisley already represented something fundamentally more transgressive and confrontational than much of the art of his time. Aside from the fact that performance itself was particularly difficult to experience again after the event, and perhaps impossible to restage, the unpleasant nature of the work made him an almost unmarketable figure (this was on the cusp of a re-emerging market ethos that would be embraced by younger artists of the later 1980s). Brisley's performances plunged spectators into uncomfortable encounters and nightmarish intimations of a world turned inside out, an empire of filth and disorder. An 'oasis of squalor' was one description of a 1970s performance that had required him to sit in a bathtub for two hours a day over two weeks, submerged in a black liquid that seemed to contain other unknown and possibly grisly items. Offal, rotting meat and other more familiar items were littered about the gallery, giving it the appearance of a crime scene in which Brisley appeared the 'victim of some disgusting, unexplained murder'.[87] Not much of it was marketable. There wasn't much left once the performances were finished – perhaps some images or video, if anyone had bothered to record him.

By the time he embarked on the project that was taking shape around the Camden and King's Cross wasteland, Brisley had begun to look at the space between the various homeless hostels as a zone of indeterminacy, an interstitial space within the mundane traffic of everyday life. It was in light of this sense of the space as one of transit that Brisley titled one of the first pieces relating to this period *Leaching Out at the Intersection* (1981).[88] It was at the intersection of the work's title – where Georgiana Street met Royal College Street – that people would surreptitiously dump varieties of rubbish, bearing out the thought that one kind of waste or wasteland, or even the waste represented by people who had been discarded by society, attracted other kinds of waste. The first item out of the dumping ground that was taken into the gallery (the Institute of Contemporary Arts/ICA) was a battered suitcase, an object that was 'redolent of passage and transition' and had 'by all accounts, come finally to rest in Georgiana Street until Brisley propelled it into the gallery'.[89]

Every day after *Leaching Out* went on show, and for the duration of its exhibition, Brisley would bring into the gallery further bags of rubbish recovered from the site – 32 in all. These pungent goods were added to what had already been accumulated, and the material would be sifted and sorted through, a daily performance undertaken while members of the public looked on. In the words of one journalist, Brisley, like 'a chef titivating the hors d'oeuvres', worked methodically through it all, vegetables, cat litter, bloodied tissues: 'There are 15 trestle-tables altogether set out in the ICA's main gallery. By the fourth day of the current performance they were quite adequately stocked with refuse; by the sixth they were getting congested.'[90] It wasn't clear what final form the amassed rubbish would take, not even to Brisley himself, who was doing something akin to 'turning a local mess into a formal, slow motion, demonstration' that, aside from its content, might rank alongside many of the artist's other epic endurance tests.[91] Eventually something like an installation began to emerge from the work of organizing the various found materials, with displays taking shape on trestle tables and on the gallery floor and walls. The recovered bags of rubbish also contained many items of discarded clothing, which would be hung from wires that ran across the gallery ceiling – makeshift airing lines for filthy clothes – and which by the end resulted in a foul-smelling spectacle, with more and more of this stuff hoisted up to the rafters to dangle in the air while being wafted around by the air conditioning.[92]

It was important to Brisley that the materials were leaching out from that dumping ground and into the space of the ICA, a place that was located in a rather more respectable environment a short walk from both Buckingham Palace and the Houses of Parliament in Westminster. Later, when it was presented in other locations as the 'Georgiana Collection', the ICA installation was supplemented with new materials and additional media components, including sound and photographs. Developed over the period 1981–6, this was conceived of as a make-believe institution whose purpose was to allow Brisley to be the curator of the rubbish he found in the small area close to his house.[93] As one of the more sympathetic of his critics wrote in a 1982 review: 'The grey images which flash onto the walls around you are soiled and simple, an old shoe, muddy piles of discarded packaging . . . When Brisley realises that the decay around him doesn't stop at the edge of the rubbish tip, so do you.'[94]

Brisley's own transition from painting and sculpture in the 1950s and '60s to performance came initially through his interest in discarded materials. In the early 1960s he was gathering materials from junkyards and 'scavenging materials from the streets and building sites of London, Munich and New York', turning a habit that was initially born out of financial necessity into a decades-long obsession.[95] Sometimes, having made sculptures from such materials, he returned his refashioned creations to the same place where he found the parts – to see if anyone would notice the difference. Apparently, they never did.[96] The Georgiana Collection later expanded to include several pieces that incorporated various elements from other works, which were reused or brought into new relationships with other materials to present new and revised installations. One piece, titled *1=66,666* (1983), presents a collection of 66 lost gloves recovered from the lost and found office of the London Underground, which Brisley filled with plaster and then painted, lending the objects 'an air of bloatedness and decay'.[97] These stuffed 'hands' were then dangled inside a steel mesh cage that was elevated off the ground and supported on two so-called 'workmates' – the brand name of a workbench used by carpenters and the like – as if severed from the now redundant bodies of those for whom the absence of work had become a reality. Each glove represented 66,666 unemployed people: when multiplied by the number of gloves, this approximated the total national unemployment level at the time.

In Brisley's work, the notion of redundancy as physical uselessness and the disposability or invisibility of the people hidden behind the unemployment figures that the news would report as evidence of the dire state of the economy together encompass a sense of waste as a phenomenon that seems to spread in certain places or among certain sections of society. Beyond that, his choice of the glove as the focal point of the piece offers another perspective on how personal possessions once held in intimate relation to the person occupy a different conceptual space than mere rubbish, making them in this case especially symbolic of how people could easily be viewed as an aspect of a homogeneous class known as 'the unemployed'. The lost glove is always, it seems, suspended between life and death, especially at the 'lost and found', where it awaits reclamation.

Just off London's Oxford Street, about a mile from the ICA, at around the same time that Brisley was collecting his rubbish, punk fashion recyclers McLaren and Westwood had created a new space to show off their latest clothing collection, where they set about creating a striking storefront and interior for a project and collection that would be known by the obscure name Nostalgia of Mud. Once remade, the new premises continued a trend they had already established by using what were ostensibly retail spaces as sites for something more akin to an art installation. Visitors entering through the front door – almost obscured by a brown clay relief of the world (after Mercator's projection) looking like something that had been thrown against the shop front – might have thought they were entering a space that could have been a leftover Second World War bomb site, the kind of place that Auerbach was painting in London at the end of the war. McLaren wanted the shop to look like 'an archaeological site under permanent excavation'.[98] The first thing that a customer had to contend with was a collapsed floor that revealed another secret 'wild' place below. Customers would then descend, as if, indeed, visiting a bomb site that contained hidden secrets, to Westwood's treasure house, where goods ostensibly for sale were spread out in front of caves along the side of the room. These openings were filled with effigies and 'jewels' and positioned around a bubbling pool of mud that was to be found in the middle of it all.[99] For McLaren, who had begun investigating Oxford Street for a film he never finished as an art student in 1969, the point was to suggest that what is pushed underground or buried eventually resurfaces: 'the real London lies underneath the pavement.'[100]

Nostalgia of Mud was something strangely echoed later in Brisley's semi-autobiographical literary work of 2003 titled *Beyond Reason: Ordure*, about a curator and another imaginary collection called the 'Collection of Ordure'. R. Y. Sirb – a partial anagram referring to Brisley himself – seeks the stuff that, in truth, never disappears, the 'objects, materials and substances which at the point of collection lie beyond the range of commodification' in some half-buried or half-hidden state of suspended animation.[101] Thus:

> Along the south bank of the Thames somewhere to the east of Gravesend there is a stretch of land which was used as a gigantic rubbish dump in the nineteenth century. The ground sinks lightly underfoot in the process of maturation. Objects rise to the surface to be reclaimed. What goes down comes up.[102]

In some of Brisley's performances, the aim of which was to be at once confrontational and disgust-inducing, he might also have been said to manifest an early incarnation of a figure that appeared a couple of decades later in the hit TV show *The X-Files*.

Writhing around in a bathtub in some abject scene or other, he was the counterpart of that show's Trashman character, a shadowy presence who identifies sufficiently with the outcasts of urban life that he conjures up a spectral avenger called the Band-Aid Nose Man, who rises out of trash cans to defend the homeless. In such a manner, Brisley's work was a sensory jolt to anyone who encountered it.

6

Temporalities: Deep, Infinite and Meaningless

The primordial condition of waste can be understood as emptiness; specifically with spaces or domains that were hostile to human existence. These connect in our present moment to the dimensions of ecology, which is also constituted as the world that waste, via its human makers, leaches out from and into.

The oldest literary sources in the Western tradition, drawing from myth and scripture, reveal that the domains of nature that were held to be incapable of supporting human life or being assimilated within human understanding – the skies above, oceans, barren and inhospitable lands – were not yet capable of being objectified in ways that extended beyond their role in supernatural belief systems. These domains were far from the humanly adapted places and environments – indeed, resources – that they have now served as for so long.

Life against the Vast

What myth and prehistory do suggest is that the human consciousness of inhospitable nature created an originary leftover in the form of the 'vast' of an undifferentiated and perplexing Creation. If we look back to some of the earliest written sources we can see the identification of something that was deemed beyond human comprehension in various attempts to understand how the human realm stood apart from, and often at the mercy of, what we must think of as the most primal aspects of nature. Written in the eighth or seventh century BC, Hesiod's *Theogony* deals with the birth of the gods, who are manifested out of the vast unknown forces that were taken as the origin point for all life. These were sometimes described in English translation as Chaos or the Chasm; first the gods emerge, and then other elements of nature:

> First of all came the Chasm; and then
> wide-bosomed Earth, the eternal safe seat of all
> the immortals who hold the heights of snowy Olympus,
> and murky Tartarus in the recesses of the wide-pathed land,
> and Love, who is fairest among the immortal gods,
> loosener of limbs, by whom all gods and all men
> find their thoughts and wise counsels overcome in their
>   breasts.
> From the Chasm came black Darkness and Night;
> and from Night came Ether and Day
> whom she conceived and bore after mingling in love with
>   Darkness.[1]

It is a vivid mythology that postulates the origins of the universe in support of a story about what made humans distinct from, yet born out of, the undifferentiated vastness of pre-creation. We might think of this infinite space today in terms of the universe, but for some of these early thinkers it was something eternal that existed as a kind of nothingness. This 'first state of matter' was termed the *apeiron* by the Presocratic philosopher Anaximander (sixth century BC) and referred to this cosmic backdrop as existing 'without boundaries' and containing within it 'an undifferentiated mass of enormous extent'.[2] This boundless universe he regarded as 'infinite in quality spatially and temporally, extending indefinitely in every direction as well as backward and forward in time'.[3] As W.K.C. Guthrie wrote in his 1967 book *The Greek Philosophers*, Presocratic philosophers of the Ionian school (such as Anaximander, Heraclitus and Thales) all grappled with the nature of this boundlessness:

> One thought of the ceaseless tossing of the sea, another of the rushing of the wind, and on the threshold of rational thought the natural explanation of their apparently self-caused movement was that they were eternally alive. Thus, we find that all of them, while in other respects avoiding the language of religion and completely discarding the anthropomorphism of their time, yet applied the name God or 'the divine' to their primary substance. So Anaximander called his *apeiron* and Anaximenes the air.[4]

It is the withdrawal from this undifferentiated natural world that defines the human, in opposition to 'the vast' and its capacious domains and powers.

This boundlessness or vastness is thus understood by reference to the human, and to human scale. And it is from the word 'vast' that the idea of waste as a human relation begins, or the Latin *vastus*, for empty or desolate, deriving important aspects of our understanding of 'waste' and, of course, 'wastelands'. The last, over time, came to refer to lands or places that had been abandoned or become uninhabitable or where human life or habitation becomes difficult. Such places and the natural forces and forms of life encountered in them sowed uncertainty in humans. They still do. From heat-induced desert mirages to the hidden depths and ferocious power of the seas, and the surprises and dangers that lay in wait in primeval forests, it is not difficult to see how they could be feared for the potential threat of death that they held. The distance between these often unstable and occasionally chaotic-seeming realms and human societies would change over time and history, to the extent that as well as being colonized more and more by human life, these oceans, uninhabited lands, deserts and skies become – by the present – receptacles for a host of new and modern wastes, from plastics to space junk, whose forms and materials were unknown to those who derived myths from uncertainty or speculated on how humanity distinguished itself from a seeming natural chaos.

As such, before waste can be thought of as a material thing that also comes to be laden with a sense of moral evaluation, it can be understood as something that describes a relation to the world, and something that refers to what either stands against the human, exists beyond it, or has been moved out of the human sphere of use or value.

Desert Infinites

'For more than half an hour the car had been steadily climbing up the winding desert road,' wrote the *Los Angeles Times* automotive correspondent L. George Thompson in a 1930 report that pitted a new vehicle – the sturdy-looking Studebaker President (in truth, it resembled a scaled-down stagecoach) – against the dry and barren climate and surfaces of the Mojave Desert.[5] For many prospectors heading west from Utah during the gold rush of 1849, it became a graveyard. These were people who had taken what probably seemed on a map to be a shortcut – a route through the desert in preference to the longer way around, which skirted the great wasteland – only to find it impossible to survive the baking temperatures of a place that would come to be known as the hottest location on the planet. Many would have unwittingly moved along routes that were known to some who made it through as the 'Jornada del Muerto' (journey of death), proceeding from one water hole to another. The first and primary definition of 'waste' given by the *Oxford English Dictionary* is 'desert, uncultivated land'. For those westward migrants, the Mojave Desert was merely an obstacle on the way to some other place. It was a vast space that might, with a bit of luck, be negotiated without calamity. In the words of one account, the path through the desert – scattered with the skeletons of horses – carried prospectors through 'large basins of deep sand', where their own mules now moved in eerie silence, the vast emptiness helping to make their journey 'more like the passage of some airy spectacle, where the actors were shadows instead of men'.[6]

A hundred and fifty years after that westward journey, a large box container was found lodged in a rocky desert outcrop; it contained personal effects, all perfectly preserved in the dry climate, that revealed the fate of one failed expedition. Included among its contents was a letter, 'tucked into a hymn book next to the verse Thirsting after God', detailing the inability of one William Robinson to go on after his last ox had perished and his travelling companions had all gone down with fever.[7] He was one among many whose shortened lives helped to name the place Death Valley. Today, there is no sign of his own mortal remains, which we can only guess have been reduced to dust and now mingled with the desert sands.

By the 1930s, almost a century on from the California gold rush, some of the problems encountered by those venturing into Death Valley and similar regions were alleviated by the arrival of vehicles like the Studebaker, which in the cooler months between October and April – when it was possible to find temperatures dipping below 38°C (100°F) – made the seemingly eternal wasteland suddenly accessible like never before. In fact, what the car revolution of that era had done was to turn inhuman landscapes like the Mojave into places that people could more easily inhabit, even if only in a temporary manner, much in the way that the eighteenth-century European Romantics had 'conquered' their own wilderness, which took the form of mountainous landscapes, and subsequently produced guides for those who would follow in their footsteps. Death Valley, no less than the Swiss Alps, was a place of extremes, as the *Los Angeles Times* informed prospective motor tourists who might experience the vantage point of Dante's View, itself a future stop on any itinerary that now made this deadly wasteland a somewhat safer place to experience. There, perched high above the valley, it was possible to take in the breathtaking view that spread out below, the dry surface stretching into the distance, where it seemed to dissolve into sky, offering a vision of a place so vast that it dwarfed the human scale of perception:

> Across the valley was a blue mountain range, snow-capped, that towered with unbroken slope from the valley floor to over 11,000 feet at the summit of Telescope Peak. At our feet the mountains fell away sharply with a drop of over 6,000 feet with what appeared to be a lake extending up to their base and spreading across the valley.[8]

The 'lake' was, however, an illusion created by the glistening salt beds and trickling streams that formed on the surface of the valley basin, and which from elevated viewpoints gave the sand the appearance of water. Who could be surprised, then, that deserts had become known to play tricks on the mind. The infamous opening lines of Hunter S. Thompson's *Fear and Loathing in Las Vegas* (1971) relay the experience of hurtling through the Mojave at great speed as a hallucinogenic vision; perhaps only partially drug-induced, as the narrator and his companion confront not only tricks played on the mind and eye, but the sense of the desert as a seemingly infinite and open space:

> We were somewhere around Barstow on the edge of the desert when the drugs began to take hold. I remember saying something like 'I feel a bit lightheaded; maybe you should

> drive . . .' And suddenly there was a terrible roar all around and the sky was full of what looked like huge bats, all swooping and screeching and diving around the car, which was going about a hundred miles an hour with the top down to Las Vegas.[9]

Even without the mind-altering substances that these travellers had taken along for the trip, the place itself was seen as the source of a distorted sense of reality: illusions stimulated by dehydration and heat exhaustion are commonplace in literary accounts of desert experiences and symbolize, at the very least, the disconnection of such an environment from routine life. 'All desert literature', David Jasper notes, 'emphasizes the mirage – the image conjured by the place that is not there yet is part of the place.'[10] The mirage represents the instability of recognizable forms under the extreme conditions found in the desert. In this, as we will see, the desert echoes something of the indeterminate or shape-shifting character of waste in its many other forms or modalities. The reversal of expectations bound up in the encounter with the desert is illustrated in the most stereotypical of these mirages: life-giving oases awaiting the one who has wandered endlessly without food or water, with lush greenery sprouting in the distance of an otherwise 'dead' landscape, or even – as in those dreams where some cherished object or circumstance evaporates on awakening – hidden treasure thought to be buried under a lone desert palm tree. 'The *mirage* has deceived us several times during the day's march,' wrote Edwin Bryant of a journey taken in 1846, his emphasis clearly on the phenomenon as something spectral or monstrous:

> When thirsting for water, we could see, sometimes to the right, sometimes to the left, and at others in front, representations of lakes and streams of running water, bordered

> by waving timber, from which a quivering evaporation was ascending and mingling with the atmosphere. But as we advanced, they would recede or fade away entirely, leaving nothing but a barren and arid desert.[11]

Bryant's journal of his passage through the desert in the years 1846–7, published in book form as *What I Saw in California* in 1849, became a guide book of sorts for gold rush prospectors heading through the desert to California.[12] In fact, such descriptions of mirages could function as allegories for what lies in wait within the phantasmagoric atmosphere and spaces of Las Vegas – a principal destination in the Mojave that stands as an example of how this particular wasteland would be transformed by human life.

More broadly, the history of encounters with desert space is informed by visions of the unreal that for all their fantastical elements nevertheless animate the looming threat that desertification poses for human life.[13] As one might expect, such uninhabited places have been represented as more likely to be 'strewn with false signals, surprises, and dangers' than they are with anything necessary to human survival.[14] It is almost a fact of their emptiness, and one that undoubtedly explains why they seem to us to remain largely abandoned or uninhabited. But beyond what is conveyed in the desert's physical characteristics, there is a conceptual understanding of barrenness in Western thought as a condition that stands in opposition to the kinds of qualities that constitute humanness or humankind. Hence, the metaphors that were deployed to reveal the very vitality of human life tended to view a particular aspect of nature as a source of vitality that supported human existence.[15] It was an association marked in works of the classical world, and one that has found its way into how we still think of human life today. Writing in the fifth century BC, the Greek poet Pindar, for instance, observed that 'human excellence' could only be understood to flourish by thinking of it in terms of plant life, 'like a vine tree, fed by the green dew, raised up . . . to the liquid sky'.[16] In such a way is human flourishing related to its proximity to nature as a garden of abundance. In the early first century, the Egyptian scholar Philo of Alexandria, in his *On the Creation*, an allegorical interpretation of the Old Testament's account in Genesis, observed that: 'Poets quite rightly are in the habit of calling earth "All-Mother," and "Fruit-bearer" and "Pandora" or "Give-all" inasmuch as she is the originating cause of existence and continuance in existence to all animals and plants alike.'[17]

Beyond its simple hostility to human life, though, the Mojave loomed large in the imagination of westward migrants as a mythical obstacle to attaining their dreams. The Great Basin, which subsumes several deserts and much of Utah and Nevada, threatened to swallow all who entered. Its status as mythical terrain rested partly on the difficulty of simply knowing or grasping the immensity of its scale. 'The older the map one consults,' Reyner Banham wrote, 'the larger the Basin figures.'[18] The uncertainty ended up making it 'the most feared barrier to the westward expansion of the United States', standing in the way of onward progress to the promised land of California.

Not long after I embarked on a trip through the Mojave Desert from Los Angeles to Las Vegas, around twenty years ago, I found myself two-thirds of the way to my destination, on Interstate 15 (famous as Hunter S. Thompson's route to the same place), looking at a sign that seemed to be confirmation that even the language of naming in this place had been exhausted by the emptiness of the environment, perhaps the lack of human signifiers. 'Zzyzx', it read, with an arrow directing drivers to an exit that looked as if it went nowhere. Looking off to the distance along the imagined sweep of this road, there seemed to be no hint of a human presence, let alone any evidence of the strangely named Zzyzx. There was only emptiness as far as the eye could see. Perhaps only the most determined scientific explorer with knowledge of the desert's sparse plant life and desert critters in such places would be able to draw on more than a limited range of referents in settling on suitable place names. In fact, the word 'dead' appears frequently in place names of the Mojave. But that is not surprising: they are descriptive names whose origins might easily be explained, in contrast to the more abstract Zzyzx. I wondered if the far more singular designation in this case had been made by some lonely prospector, who believed that they had reached a point of no return and decided that this place should be named something other than Dead Man's Flats or Devil's Plain (or countless other names that signify death).

Writing now, the thought brings to mind the observation made by Daniel Heller-Roazen that, when a language seems to have run its course, 'the name we are accustomed to [giving] such an end is the one we use in reference to organic being: death.'[19] Could Zzyzx have been one of the many ghost towns that were dotted throughout the Mojave and acquired its peculiar name

The road to Zzyzx, signposted on Interstate 15, California.

as a result; one of those places founded during earlier westward migrations or expeditions to locate minerals or other valuable natural resources? It wouldn't have been unusual, because the 'failed cities' of this desert, as the film critic and historian David Thomson has noted, seem to 'outnumber' those that succeeded.[20]

I later discovered that Zzyzx was not in fact one of these ghost towns, and that the road off I-15, which today leads to the California State University Desert Studies Center, had once arrived at the destination of Zzyzx Mineral Springs and Health Spa, previously known more simply as Soda Springs. Or, as the authors of a travel companion titled *Weird California* note, this strange road delivered one to 'a Utopia spelled Z, Z, Y, Z, X'.[21] Despite its invisibility from the road sign in question, Soda Springs was a destination just a few miles along a bumpy sand track off the interstate, the details of which – before mapping technologies became easily accessible through mobile and portable devices, such as sat nav and smartphones – would mostly only be found in off-beat guides like *Weird California*, or hard-to-find books like Reyner Banham's *Scenes in America Deserta* (1982). Banham, indeed, describes its location as 'beautiful: a small promontory on the eastern "shore" of Great Soda Lake at a point where there is fresh water' and 'one of the greatest resort sites in the whole of America Deserta', by which he means the deserts of New Mexico, California and Nevada. It was, he noted, one identified by a road sign that most people take to be a 'typographical error'.[22]

As Zzyzx, the settlement had been founded in the 1940s by Curtis Howe Springer, a health freak and radio preacher who took up residence there at a time when the desert was still largely unoccupied, and who 'touted the medicinal properties of the mud from Soda Springs'.[23] 'To radio listeners in 50 states and scores of foreign countries', it was reported in 1967, 'Dr Springer is a friendly philosopher, a disc jockey who plays religious records and quotes from the Bible on daily half hour programs as he peddles a line of health foods and promotes his resort of Zzyzx.'[24] Springer had built the resort after simply assuming that this was a wasteland that no one had any prior claim over. Some years later a *Los Angeles Times* report dubbed him a 'super squatter' after it was discovered that he had laid out 'a complete town on federal land to which he [had] no claim'.[25] He was eventually targeted by federal authorities, who declared ownership of the land and closed the resort down in 1974.

The origin of Zzyzx, as the meeting place for a corporeal spirituality induced by the magic mud found at the springs, is perhaps not surprising when we consider that the settlement was established on the California side of the Mojave. The early history of that state – since its settlement by the westward migration that eventually overcame the obstacle that the desert represented – was significantly shaped by a motley assortment of health cranks and religious nonconformists who manifested their separation from the old and the unwanted in a proliferation of unconventional systems of belief – variously spiritual and lifestyle-related – which have continued to flourish in California.[26] At some time in the development of a California identity that mixes Eastern mysticism with frontier individualism, spirituality and health were conflated, producing a new religion of the body beautiful.[27] One of California's chief attractions to migrants was the supposedly health-giving virtues of the land and climate. This usually referred to its proximity to the ocean and its abundant fruit orchards, but in the desert – and this was probably what made the site of Zzyzx uniquely attractive – the stripping away of routine life marked it already as a quasi-spiritual zone where 'intoxications of sight' and 'exaltations of perception' were reportedly commonplace.[28] Come to the desert to get out of your mind, seemed to be the message.

In longer historical terms, a host of religions had developed an intimate connection to the desert precisely because the reversal of human comforts that could be experienced in the barren surroundings provided grounds for the ultimate test that could concentrate the mind on other than earthly matters, elevating the spiritual above mere material life. It is no coincidence, then, that the world's major monotheistic religions are also known as the desert religions, and they are seen to have emerged from places that were able to develop narratives about the role the desert served as a site where God tested things out 'before making the rest of the world'.[29] Early studies of Christianity, in particular, reveal deserts – and other wild and remote places – to have been perfect locations for the withdrawal and denial of bodily needs that would come to be demanded by monastic ideals, because it was there that God was most likely to be found, and where people pursuing some ascetic ideal 'encounter the inner and outer demons'.[30]

Today, as already noted, the deserts of the American West are littered with the remains of failed human settlements that date from the gold rush era as well as later periods, their crumbling remains the material evidence of human limits. By contrast, the natural features of these landscapes appear to exist in a condition of timeless immutability, unchanging, in what seems to be a dimension far beyond the temporal limitations of human societies and their relentless drive to remake the world. People are nonetheless still drawn to the desert as a place where human limits can be tested at various locations, including the city of Las Vegas and the annual Burning Man festival. There is a continuity between the experiences that can be had in those places and the idea that the Western frontier – and the destination beyond, California, which represented a new beginning – could only be reached by enduring a trial in the desert.

But if human history was shallow in the American desert, the human encounter with its landscapes soon revealed that it was a place marked and shaped by natural forces that mostly lay dormant but at the same time pointed towards another, non-human, temporality. Geologists now think that what the long future has

Burning Man festival, Nevada, 2014.

in store for this place begins at a level far below what is observable on the surface, with tectonic plates shifting and pulling the land mass of California apart from the desert, creating an opening that would allow a large body of water – a sea – to establish itself in the Mojave, cutting off parts of present-day California from Nevada. This could be the first stage in a new geological phase that might, moreover, see what is known as California split off from the rest of the continental United States.[31] 'California will be an island,' one geologist confidently told the writer John McPhee, 'it's just a matter of time.'[32]

But for now, what exists on the other side of the area that may one day be filled with enough water to constitute a new inland sea are places and landscape features left behind by previous geological upheavals. In pioneering photographer Edward Weston's 1930s images of Death Valley, the strange and beautiful forms of the Badlands canyons at Zabriskie Point, whose craggy forms today sit beneath a tourist viewpoint, look little different from countless pictures of the same landscape taken almost a century later, in

'From Wasteland to Wonderland' reads the caption on an information board at Zabriskie Point in Death Valley, California.

our own time, by tourists. The latter now have easy access to the desert and the means to move around it in the relative comfort of air-conditioned transport and with portable equipment able to generate limitless images of what remains a deadly place. But, as the proliferation of warning signs shows, the conveniences of modern mobility can lull the visitor into the false sense of security offered by the impression that this is a land that has finally been tamed.

Things were a little different in Weston's day. After winning a Guggenheim Fellowship to document the landscapes of the region, he felt it necessary to spend two years doing so (a feat only achieved with heavy and cumbersome equipment and the help of an assistant). He accumulated approximately 1,300 negatives (a number that a keen amateur now could match in digital images much more quickly), taking roads less travelled and barely existent as he clocked up some 35,400 kilometres (22,000 mi.) to obtain pictures that might illustrate how modestly we humans stood against the deep time of the desert.[33] The figures we have on the number of negatives made and distance travelled by Weston vary depending on the source one consults: those just quoted are complemented by others that put the figure at 1,500 negatives and 56,300 kilometres (35,000 mi.) travelled.[34] Even these larger figures seem modest within a contemporary context. Nonetheless, what such images of the desert – both those from the last century (and even earlier) and those made by others today – illustrate, when taken together, is the permanence of the landscape and its apparent indifference to human intervention and inhabitation. On one level, however, this is an illusion. Studies of the desert also show human life as having had more long-lasting effects here, in the form of the new and alien plant species that were brought from foreign environments. The salt cedar shrub, for instance, is a plant native to Asia and Europe, but one that in the Mojave disrupted the balance of the local ecology, sucking up 'subterranean water sources' and 'exterminating wetland habitats previously occupied by native plants and animals'.[35]

Into Oblivion

From the vantage point of the Mojave Desert that he lived in during the 1940s, English writer Aldous Huxley was awed by how quiet and infinite the landscape around him seemed; here was an area of land as large as Europe, the ground of a seemingly endless vault that extended up to the heavens. It was a space that could swallow up the sudden, violent appearance of the most ear-splitting of the occasional invaders overhead. Right then, at the start of the age of the jet engine, he watched planes burst into the skies above, their impact immediate but soon enveloped again in silence, as if to suggest that this was a place that would preserve no memory of the temporary human presence of which he was a representative: 'the screaming crash mounts to its intolerable climax and fades again, mounts as another of the monsters rips through the air, and once more diminishes and is gone.'[36]

If it could be said that it is in the many natural features of the world's desert regions that we are able to distinguish or identify what is unique to them, it it is equally true that the nature of the desert as a particular kind of wasteland, or an adapted wasteland, can only be understood through the human relation to it. The very notion of wasteland is, in this context, inextricably bound up with the perception of the desert as an unremittingly hostile place that is not particularly suited to human inhabitation. And just as all understandings of 'waste' are more generally meaningless beyond some human perspective or system of values, so too the place of the desert in the human imagination and as a cultural concept tells us more about ourselves than it does about the desert as a natural environment. For what the desert has always ultimately referred to, in varying degrees, is the absence of the human – that is to say, it points to the unpopulated, the abandoned and the inhospitable as seen from our perspective.

And because the desert – like waste 'itself' (in its many manifestations) – exists at the limit of human experience, it seems reasonable to agree with Reyner Banham, who declares in *Scenes in America Deserta* that 'the ultimate definition of a desert may yet prove to be concerned with the number and type of people present, and what they think they are doing there.'[37] In other words, is the desert *as* desert, after all, another one of those human products that represent the extension of our desires and projects? I might go further and say that in its relation to the human, and specifically as a peculiar kind of wasteland, deserts are – or can quickly become – the kinds of spaces or places where the radical contingency of human existence, its dependence on almost everything that the desert in its natural state lacks, becomes apparent. The disruption of normal experience allows the desert to appear as a space of forgetting, and one where waste, as a phenomenon that exceeds human limits, in a sense finds its place.

The Mojave Desert began to grow as a place of human habitation during the Second World War, a trend that continued into the Cold War era, drawing in all kinds of people who saw it as a space in which to expand their own potential: 'from millenarian cultists to visionary artists to the advanced weapons scientists of the United States Air Force'.[38] Desert research centres cropped up to study the geology, phenology and vegetation particular to the region. More recently, human population growth in desert cities like Las Vegas and the activities of the U.S. government and military on the desert environment are extending human life and culture even further into the region. Las Vegas began to grow in the 1940s with the arrival of government-funded defence contractors. Traffic flowed between California, the location of munitions manufacture, and the town of Henderson, southeast of Las Vegas, where magnesium was produced for the defence industry. By 1947, Highway 91 – a section of which would become the Las Vegas Strip – pulled in a broader range of tourists from Southern California, expanding the potential for resort hotels.[39] A peculiar human geography has ensured the desert's emergence as another form of waste space: a destination for activities and entertainments that were designed as a means to overturn the norms and expectations of everyday life.

The reason I never ventured off the freeway to the mysterious Zzyzx was that I still had some way to go before reaching my destination – Las Vegas. It is, of course, the foremost of the many desert playgrounds in this vast landscape, all sites of conspicuous consumption, excess and waste. Las Vegas draws in visitors from all over the world today, but in earlier days its entertainments were pitched primarily at weekend revellers from Los Angeles, making that city to all intents and purposes the centre of Vegas's 'Californian hinterland'.[40] There, attractions at the desert resort were often advertised on huge billboards along Hollywood's Sunset Strip, beside other such signs for entertainments more local, giving the impression to the unwise that Vegas was located in a nearby locale rather than nearly 315 kilometres (195 mi.) away and accessed through a landscape that one would not want a vehicle to break down in.

Casting myself in the position of a tourist set on reaching the other Strip in Las Vegas, I had put thoughts of the distance between the two cities to one side as I, like countless other travellers, set out from Santa Monica, the terminus of the old Route 66 by the Pacific Ocean. From there it wasn't difficult to negotiate oneself towards the on-ramp of Interstate 10, the San Bernardino Freeway, which leads directly to Interstate 15, linking the two cities in a way suggestive of them being the twin poles of American excess.

Once advanced beyond the Zzyzx road exit, the next significant point of divergence on my route was headed to Death Valley, which, even in 1982, in the view of Banham, was approaching a Disneyfication of over-regulation and prescribed visitor attractions. It had been 'just about labeled to to death', he noted, because of 'a sense of constantly being told what to think, how to react, as if the National Park Service dare not leave people to employ their own perceptions and come to their own conclusions'.[41] According to other observers, the status of the Death Valley National Park amid the surrounding no-go regions of the desert – which are correspondingly cast as de facto wastelands – reveals something of the politically charged history of this place. Valerie Kuletz argues that the labelling of pockets of the desert as either 'wasteland or national treasure demonstrates the arbitrary nature' of the way that such descriptions are applied, but also, on the other hand, how they can be 'strategically used for political ends'.[42] In other words, perfect for concealing things that politicians or other interested parties want hidden from the public.

Yet Death Valley's location in the vast spaces of the desert along with its extreme climate – where temperatures can reach

54°C (130°F) – continues to lure others in search of extreme experiences, where they will be met today with warning signs advising that walking after 10 a.m. presents a serious risk to life. It also provided a strange backdrop to the notorious hippie 'love-in' sequence in Michelangelo Antonioni's 1970 film *Zabriskie Point.* 'Repellent', opined one review of the film; 'It offers, on the wide screen, a sweeping view of our Western desert, populated with naked couples, scores of them all making love there on the burning sands of Death Valley.'[43] When he first met with executives from the studio, Antonioni said that he imagined 'ten thousand people making love across the desert'.[44] In the end the number was more like a few hundred, with extras from Las Vegas coming onto a set the director thought to be like the surface of the moon. Although the scene, as scripted, was considered scandalous before it was even shot – and the park rangers who facilitated access to Death Valley were dead-set against it, believing it would bring the wrong kind of people to the place – the studio were happy for it to go ahead, thinking that it would guarantee no shortage of free publicity for the film.[45] The love-in scene was another of those desert mirages, with lead character Daria – after smoking pot – having a vision of 'sand-colored forms rising from the earth who embrace, frolic, and copulate in myriad configurations'.[46] Given what else went on in this desert, the film may have done little more than become one more cultural reference point for the idea of the desert as the site of inverted moral values.

The idea of the desert as a space into which one can easily vanish without leaving much of a trace brought the notorious Charles Manson and his 'Family' here in September 1969, to a location known as Barker Ranch, after they had committed the series of murders in Los Angeles that would finally see them captured. Manson accumulated followers among social outcasts, transforming himself within a short space of time from a 'homeless parolee to leader of what was, in essence, a personality cult'.[47]

Preaching selectively from the Book of Revelation and embedding the use of psychedelic drugs as part of 'the Family's working rhythm and internal group dynamic', he convinced his followers that 'the End of Days [was] a great and imminent reckoning.'[48] This he referred to as Helter Skelter, after the Beatles song of the same title. Manson was convinced that the song was commanding him to unleash a race war that would bring on the final reckoning, the End of Days of the Book of Revelation. When he arrived in Death Valley with some of his closest followers to make preparations for their escape, they came heavily armed for a final showdown with the authorities. Others 'arrived the following week, driving stolen dune buggies and other vehicles [and] began setting up lookout posts and fortifications, hiding caches of guns, gasoline, and supplies'.[49]

Much like an evil counterpart of the desert mystics of early Christianity – who went into the desert in search of a revelation from above – Manson believed himself to be within reach of a route underground to some imagined paradise.[50] Once he had unleashed his planned 'Helter Skelter' revolution, creating chaos in America, he aimed to lead his followers to a cave that was the entrance to 'a golden city' that awaited them, as one of his followers later revealed:

> there's a river that runs through it of milk and honey, and a tree that bears twelve kinds of fruit, a different fruit each month, or something like that, and you don't need to bring candles nor any flashlights down there. He says it will be all lit up because the walls will glow and it won't be cold and it won't be too hot. There will be warm springs and fresh water, and people are already down there waiting for him.[51]

In choosing the desert as the most fitting location to disappear, Manson – consciously or otherwise – made a symbolic choice that followed in a tradition of outsiders and outlaws who had also been drawn to places that were essentially lawless or difficult to police. But Manson also painted himself and his followers as just a product of the society that would condemn him, even suggesting that they were just another of the inevitable wastes that human society created. They represented something that is perhaps kept down or under control, but always ready to breach human limits. As Manson himself declared in one of his court appearances:

> You can jump up and scream 'guilty,' and you can say what a devil fiend, slimy devil I am. I am whatever you make me . . . Where does the garbage go, as we have tin cans and garbage alongside the road, and oil slicks in the water, so you have people, and I am one of your garbage people.[52]

Nevada Test Site

The Nevada Test Site (NTS), a huge area covering 3,500 square kilometres (1,350 sq. mi.) of desert northwest of Las Vegas, was just one of several such locations established in the United States during the Cold War by what was then called the Atomic Energy Commission. It was formed in 1946 with a programme that, by 1953, was being described within the terms of President Eisenhower's 'atoms for peace' policy. The vast area of the NTS reconfigures the desert as no ordinary dumping ground, but rather a far more expansive 'technoscientific' wasteland.[53] This makes the NTS an area of different dimensions to those other playgrounds that can be found in other parts of the desert – and which are in their own way wastelands: Burning Man, Zabriskie Point, the gambling cities – because of the nature of the human communities that have come to inhabit it. The first atomic bomb test occurred in another place, at the Trinity site in the New Mexico desert (on 16 July 1945), following which test sites were opened in the Mojave. Since then this desert region has been home to 'the most dangerous projects of an industrial, militarized society'.[54] Tourists from Las Vegas, indeed, were able to partake in aspects of the culture that grew out of such projects, entertained 'not only by above- and below-ground nuclear tests, but also by spectacles such as the Miss Atomic Bomb pageant and refreshments such as the atomic cocktail'.[55]

But for much of its early history the NTS was not to be found on maps of the desert. Today it is possible to identify its boundaries, which actually lie within the much larger area known as Nellis Air Force Base, a site that itself is home to a wide range of military training activities and scientific experimentation. It was also within the Nellis base (in the mysterious Area 51, the source of conspiracy theories about alien invasion cover-ups that have spiralled out into the popular consciousness through film and television shows such as *The X-Files*) that UFO sightings became commonplace during the Cold War, possibly involving prototype military aircraft that were being tested in the desert. For those who believe in the cover-up, in the notion that this is the place where alien spacecraft are hidden, it was here 'that military scientists used reverse engineering to deconstruct' such a craft, 'after it crashed in Roswell, New Mexico in 1947', in order that we could learn how to apply 'alien technological magic in our own weapons'.[56]

That there are several distinct sites within this vast area of the Mojave gives a hint not only of the quantity of space given over to military-scientific experimentation, but how easily it is all swallowed up within the capacious void of the desert. To look at it in comparative terms, this is a region that is roughly the size of Switzerland: a small nation, to be sure, but one that is home to some 8 million inhabitants. Valerie Kuletz notes that the military presence in the Nevada part of the desert alone, which is dispersed across a variety of locations, covers some 'four million acres'.[57] On the California side of the desert, one might locate Edwards Air Force Base, or try to find the highly secret, long off-the-map China Lake, 'one of the largest weapons research and testing facilities in the world', where scientists and technicians, securely and out of sight, 'work hidden behind mountains'.[58]

And in addition to the abandoned towns and cities of the desert, there were once simulated settlements built by the U.S. government as part of the weapons testing programme. The model town known as Apple II, for example, played its part in nuclear bomb experiments, allowing experts to examine what might happen to people in their homes in the event of nuclear catastrophe. In the testing of weapons, it was necessary to gauge the impact of the deadliest kinds available. A series of typical structures acting as stand-ins for run-of-the-mill everyday buildings – a school, a library, a bank, and so on – were erected. This was in addition to newly built stand-ins for human dwellings: 'a dozen homes were built and furnished with everyday items (televisions, refrigerators, furniture, carpets, and linens). They were then stocked with food, populated with white-skinned mannequins, and neatly incinerated.'[59] Photographs taken during the experiments, which can now be found in the U.S. government's own Library of Congress picture archive, show mannequins pre- and post-test: a female figure surrounded by three 'children'; a family at rest in their sitting room; another family sheltering in the basement. Later they can be seen collapsed amid the wreckage of blown-out buildings.

Aside from the damage to property and the implied visitation of death on any human presence, the secrecy surrounding such testing all seemed in keeping with the desert's capacity to hide or absorb the effects of anything that the human desire for tinkering with nature could come up with. This is a fact confirmed by military communiqués that circulated in the planning stages of the NTS, which describe the site as a place that the rest of the country remained unaware of, making it perfect for the purposes that the government had in mind. It was of little use for anything but 'gunnery practice', the military claimed, meaning that it could be bombed 'into oblivion' and no one would know any better.[60]

In fact, the activity was driven further into darkness in the early 1960s when weapons testing went underground, thereby seeming to vanish from even the limited public awareness that accompanied earlier testing, which had at least been visible in the skies many miles away. 'The desert was the perfect home for the Cold War,' notes Tom Vanderbilt in *Survival City*, a 2002 study of so-called Atomic Towns, 'the last America, the last New Frontier, where dreams might be played out with a minimum of interference.'[61]

The last nuclear test that took place here was in 1992. What remains is a landscape transformed by 'nuclear explosions that melted the bedrock and fused the sand', where 'hundreds of moonlike craters dimple the wasteland.'[62] Today, signs around the perimeter of the NTS declare its use for 'the testing of devices for defense and for peaceful uses of nuclear explosives'. Such language further plays up to the 'environmental research park' image that the authorities project by also alluding to the land's prior use by 'prehistoric, aboriginal, cultures', as if what it is now – and has been for decades – forms part of some natural evolution or continuity in human society (perhaps it does) and is an example of the military's affinity with the land. But these signs, of course, obscure the nature of this wasteland. To look beyond the obfuscations of the official language surrounding the activities that have taken place here, Mike Davis suggests, is to find a 'landscape of incomprehensible devastation, sown with live ammo and unexploded warheads'.[63] As others have pointed out, the NTS now contains well over a thousand contaminated sites that require cleaning up, creating a situation where it is feared radiation might leak into the groundwater and 'slowly move towards Las Vegas', the greatest concentration of human population in the desert.[64]

The relative obscurity of what goes on in this landscape and its 'secret science cities' – declared off-limits to the public – also, of course, deflects attention from the consequences of the activities that take place here. But the perception of these Mojave test sites as wastelands is not merely evident through encounters with the trash of nuclear experiments. It is rather an awareness that is raised by knowledge of the strategies required to come to terms with the by-products of not only nuclear weapons testing, but spent nuclear fuel, especially insofar as it was expected to find a destination here in the desert. In this case, the everyday equation often applied to our logic of waste disposal – out of sight, out of mind – speaks of a more alarming dilemma. Namely, material that will remain dangerously toxic for periods estimated on such a timescale that it escapes the scope and comprehension of any model previously devised for the study of human culture invites a new kind of attentiveness: one quite different than how societies have approached the polluting effects of more mundane forms of waste, which, in most cases, can be kept within the scope of a temporality that is of a human scale, making the management of them routine. The attention required by the problem of nuclear waste involves a revolution in how we devise ways of shaping memory as it relates to place.

Clearly, the absence of a meaningful temporal horizon for simply grasping the idea of the toxic *longevity* of the most dangerous nuclear waste means that the gravest danger lies in the fact that, once it has been subjected to the preferred method of containment, deep geological burial, its existence and threat may – over periods of thousands of years – be easily forgotten about. It is the stuff of sci-fi visions of the future, dating back as far as Wells's *The Time Machine*, where the population in some distant future are seen to have forgotten what their ancestors knew. The desert seems to add complexity to the equation because it represents an existential void, an essentially non-human landscape.

And just as the military and scientific activities that take place are easily concealed in the immensity of the desert, so too might any place marker intended to communicate a sense of unresolved danger to inhabitants of the distant future be swallowed by that unknown temporal horizon.

The specific practical danger is that a toxic dump, such as that which had been planned for the Yucca Mountain site that lies partly within the NTS (and which was put on hold by the U.S. government even before it was due to open in the 2010s), might be accidentally rediscovered and disturbed at some unknown time in the future. It is a possibility that has invigorated attempts to devise a durable monument to the remembrance of what might one day lie buried here deep below the ground. Some of the designs for suitable monuments are reproduced in Peter van Wyck's illuminating 2004 book *Signs of Danger: Waste, Trauma, and Nuclear Threat*.[65]

Part of Operation Upshot-Knothole was a 15-kiloton test fired from a 280-mm cannon on 25 May 1953 at the Nevada Proving Grounds (now known as the Nevada Test Site), Frenchman Flat, Nevada.

Looking at the Waste Isolation Pilot Plant (WIPP) site, located in the New Mexico desert, Van Wyck reveals that the concern was not only with creating an overground 'sign of danger' that might persist in something more akin to a 'geological' timescale (which is to say, tens or hundreds of thousands of years). It was also felt important that such a sign might be read by, or could communicate a message to, a future people who may – let's say, 50,000 years hence – have no knowledge of present-day languages, or access to the ready-to-hand symbolism that seems universal in human societies today, to identify a site or location that contains grave dangers (such as a skull and crossbones). The sci-fi dimensions of the scenario are equalled by the fact that contingency plans have gone so far as to imagine a future in which extraterrestrial visitors might one day move around in such polluted sites both unaware of what has gone on before and oblivious to the dangers inherent in disturbing any site where nuclear waste is buried. In other words (and in some sense to connect this logic to conspiracy theories about prior alien abductions), the threat to humanity from the (apparently safe) deep burial of toxic materials being disturbed is acute enough that hypothetical outer space aliens, too, would have to be warned, lest they inadvertently release the danger.

Las Vegas

A different oblivion is most easily found some distance from the NTS, in the experience found in Las Vegas. Even on the surface, before one deposits a single coin in a slot machine, the city offers up a vision of excess: 'a radiant spectacle of burning chrome, sparkling glass, and towering neon signs'.[66] This is the Las Vegas everyone is familiar with; the glittering bauble that lures in those willing to throw away wealth that is otherwise hard-earned. In the words of Mike Davis, it might be hard to avoid seeing 'too many people in the wrong place, celebrating waste as a way of life'.[67]

If that referred to what was thrown on to the gambling tables, then that was merely another kind of waste, but produced and particularly exaggerated in this setting. As John Gregory Dunne discovered when he skirted the edges of the city, the waste that was evident in gambling did not supplant the regular kinds of wastes found elsewhere, or back in 'normal' society. Rather, it supplemented it. Thus, as the Vegas Strip petered out and his vehicle approached the emptiness of the desert, what Dunne found was a road littered with miscellaneous unwanted leftovers, the stuff of the everyday lives that continued behind the glitzy facade:

> carcasses of cars, refrigerators, propane heaters, furniture with the stuffing ripped out, a dump that was not even an official dump. Just a place in the desert to dump the leavings of a lifetime . . . Tires, old radios, television sets with no picture tubes, stoves, washing machines, bicycles, ironing boards, supermarket carts, air-conditioning units.[68]

And so on. In a strange synchronicity, what he found illustrated a perception of place that accorded with the earlier assessment of the U.S. military: namely that the desert was exactly the sort of place that could serve as a site that swallowed up as much chaos and destruction as you wanted to throw at it. Probably the most important factor in making Las Vegas the popular destination it became, in fact, was its location in a desert, with the literal wasteland of its environment providing the necessary contrast between scarcity and abundance that lent it the veneer of the ultimate mirage: a place of luxury, the 'oasis-to-end-all-oases'.[69] Being able to fully appreciate this contrast and marvel at the excess that drives life in the city depends in part on how the destination is reached, as a host of witnesses have attested.[70] Arriving via road from Southern California offers its own experience. 'Once you get sixty miles out of LA and into the high desert,' writes William Fox, 'your senses are loosened from reality.'[71] From the perspective of a driver, the interstate seems like an endless line extending ahead on what might as well be the surface of some distant and uninhabited planet. The desert trip can thus acquire a mirage-like power to suck the traveller into a different experiential space, one whose disorientating effects were undoubtedly more intensely felt in the days before vehicles had air conditioning (regular cars going back to the earliest days of desert adventurism offered no protection from heat exhaustion).

My own sense of accelerating into some strange liminal zone during my one and only trip along this route was accentuated by a sudden awareness of the gigantic roadside billboards that punctuated the endless desert plain, signalling the proximity of some minor distraction that was illegal on the side of the border I was just about to leave, but about to be imminently available should I be tempted, as soon as I crossed into Nevada. I was within one hour of the state line, which meant that the signs I began to see were also essentially marking time – specifically in the sense that they acted as a countdown to potential transcendence through the quick and easy abandonment of the temporal strictures of routine life. One sign, displaying the whirring barrel and numbers of a gambling slot machine, blazed into view, teasing with the words 'Ready to Spin?', and was then quickly behind me. On the road ahead, not far after that gigantic billboard had zoomed into view, was Whiskey Pete's, the first structure on the road to Las Vegas where gambling was the main economic and leisure activity. A lone hotel-casino in the middle of nowhere, ready to cater to those unable to hold on until the gambling capital of the world itself could be reached. But even taking the quicker route by air, a visitor will find that they emerge into a similar teaser for the main action: the airport arrival resembled the floors of the city's casinos, with slot machines all over the place and with the sights and sounds of lights and dinging bells, chatter, and the rush of cascading coins battering the senses.

For Jean Baudrillard, who was perhaps more used to imagining casinos within a European context (places like Monaco, where gambling was locationally proximate to normal life), the affinity between gambling and the desert seemed 'mysterious'; yet on another level – that of waste – it made perfect sense. It represented what was obviously a perfect accommodation of human and natural environments, which didn't so much collide as oppositional forces as merge seamlessly, two kinds of waste constituting a singular and unique wasteland. The 'sublime natural phenomenon' presented by the vast emptiness of the surrounding landscape and the hostility of the climate was matched by 'the abject cultural phenomenon' of the city of Las Vegas, a place without limits. Las Vegas appeared limitless whether one identified its 'no limit' infinities in terms of the way that the city plundered the wealth of those who flocked to its casinos, or as an aspect of its voracious use of the region's natural resources, including its finite water supplies, to maintain the illusion of abundance in such desert surroundings.[72]

The submission to chance that defines the gambling experience represents an engagement with an apparently infinite and uncontrollable force. It explodes human limits and, as such, becomes symbolic of the existential reversal that is death. Perhaps it was something that was always known by those who gambled with their lives to carve out a life in this desert, if one takes as evidence the prevalence of certain place names that seem intended as memento mori: Dead Mountain, Dead Horse Point and Deadman's Canyon, among others, contain such hints, as do depictions of gambling and gamblers at the absolute limit in Las Vegas that can be found in film and literature, a favourite of which – by English writer Jonathan Rendall – wraps such ideas up in its title: *This Bloody Mary Is the Last Thing I Own* (1997).[73]

Where Baudrillard in his rumination on the city and its desert setting was preoccupied with the metaphysical, the California native Mike Davis looked at Las Vegas and wondered what could have motivated those who planted the city in such an unlikely location. The answer soon became clear. Where else would you put such a 'strange amalgam of boomtown, world's fair, and highway robbery' except in an environment that would seem unable to support it – a stage set – but one that could be regularly reinvented or remade on the surface to maintain its glittering illusion for generation after generation.[74] It is therefore a fantasy world that conceals many anomalies, a night-time milieu where darkness cloaks the hidden, backstage workings. No water in the desert? That's just a flaw in the reality, which has now been left behind. In the sweltering heat of Las Vegas, a fine spray of water may accompany one's passage from a sidewalk onto a moving walkway, carrying the crowds into the air-conditioned, temperaturecontrolled gaming rooms of tastelessly opulent casinos. Today's mega-casinos, departing from the simpler Western stylings of the early days, present fantastical illusions that reinforce the mirage-like quality of the Vegas experience. They have featured at times simulations of a lush Polynesian abundance, a mirage that has to be maintained by an unsustainable 'hydro-fetishism' that sees extravagant fountains and water installations in front of the gambling rooms, while hotel suites rival the fantastical architectural forms of the cityscape as spectacle.[75] This was one of the many examples of 'Junkspace' identified by Dutch architect and theorist Rem Koolhaas in its exhibitions of 'denaturing':

> a benign ecofascism that positions a rare surviving Siberian tiger in a forest of slot machines, near Armani, amid a twisted arboreal Baroque . . . Outside, between the casinos, fountains project entire Stalinist buildings of liquid, ejaculated in a split second, hovering momentarily, then withdrawing with an amnesiac competency . . . Air, water, wood: All are enhanced to produce Hyperecology™, a parallel Walden, a new rainforest.[76]

If such abundance is an invention, nothing less than a fiction made real in a place where nature provides low rainfall and infertile soil, the status of Las Vegas as – in the words of one recent study – 'the least sustainable city in the nation' is reinforced by the fact that the 'whole economy of the city' relies on the steady flow of tourists, a resource that, in an era of climate change anxieties around long-distance travel, could quickly and easily disappear.[77] In the meantime, let us not forget the other postmodern mirages within this cityscape of weird simulacra, as found in the miniaturized copies of parts of iconic cities: here, a bit of Manhattan or Paris; there, the Venice canals and gondolas, under an artificial sky. Las Vegas provides the means to subvert all kinds of identity relations that enable us to firmly locate ourselves within time- and place-bound reality. From the gaudy hotels and bars of Fremont Street and the Strip to the drive-in wedding chapels and the spectacular, illuminated facades of night-time, here one might justifiably believe one has been temporarily marooned in a dream.

Fireworks light up the Stardust hotel in Las Vegas, Nevada, as demolition charges inside begin to take down the building, 13 March 2007.

The city's relationship to the real is perhaps best summed up by the impermanence of appearances at almost every level. These range from the manufactured natural environment to the seemingly temporary front presented by the glitzy facades of its main streets, which are regularly torn down to be replaced by newer and more alluring distractions – obsolescence is a key fact of the city's cultural existence, where 'extravagant demolitions' of now stale or redundant-looking casinos 'have become Las Vegas's version of civic festivals'.[78] Even what we would term Vegas's 'natural' features are temporary stand-ins, with trees uprooted and moved if necessary to conjure a new mirage to keep the customers coming.[79] The city's ambience of existing outside of time, of course, is part of the appeal and a source of its ability to make anyone divest themselves of their normal limits. For some, however, the city now plays a 'game of effective history', developing its own relation to its trashed pasts and abandoned structures by assimilating elements of bygone Vegas into the tourist economy.

This is seen in attractions such as the YESCO company's Neon Museum. One might suggest that the dismantled, yet iconic, signs of old attractions, scattered indiscriminately across this 'boneyard', attain a poetic reversal that tugs at the seemingly insatiable contemporary craving for nostalgia, creating a 'mythologised collective memory [that] plays with the image of Las Vegas as a place of constant change'.[80] For architects Robert Venturi, Denise Scott Brown and Steven Izenour in their canonical text *Learning from Las Vegas* (1972), the changing surfaces of the cityscape, particularly along the Strip (where neon had taken over from earlier glitter-style exterior decor that reflected back ambient or natural light), could be taken as representative of the speed of life. 'The mechanical movement of neon lights is quicker than mosaic glitter,' they observed:

> the intensity of light on the Strip as well as the tempo of its movement is greater to accommodate the greater spaces,

Discarded signs at the YESCO neon junkyard in Las Vegas, Nevada, c. 2006.

> greater speeds, and greater impacts that our technology permits and our sensibilities respond to. Also, the tempo of our economy encourages that changeable and disposable environmental decoration known as advertising art.[81]

Before it attained its current status as the fastest-growing city in the western United States, Las Vegas had long been known as a unique kind of nowhere, more often a temporary destination than a place chosen to live. This is unsurprising in the sense that the desert presented a challenging environment for the human inhabitant due to the absence of water and electricity and the other necessities of life. But as a nowhere destination, it drew in people in search 'of what they and their world might yet be' in ways that served only to validate the city's mythical power of transformation. The array of attractions – the 'entertainment, money, sex, escape, deliverance, another chance, a last chance . . . another life for a few hours, days, forever' – all create the illusion of a quick transformation, the easy disposal of the weight of daily cares, if not the past.[82]

Perhaps the most revealing evidence of how deeply the dream of desert transformation penetrated the popular imagination is found in the example of Elvis Presley. Largely eclipsed in the 1960s by a new generation of performers, Presley was given one more chance in Las Vegas, but in the end the city would consume him, just as it does most of those who stay too long. Unsurprisingly for one stuck in such a fabled dreamworld, the once great icon of the twentieth century ended up as an exaggerated and white-suited cabaret ghost of what we have to accept was the original or 'authentic' Elvis. In Vegas, he embodied a vision of tasteless excess, dressed as if to match the gaudy character of the place in glittering stage suits and capes. His last years were perhaps also tied up with the gambling debts his manager had been accumulating for years since the two had arrived in the city in the late 1960s. It was a situation that has been said to have seen Elvis transformed one more time into a deluxe gambling chip: a star to be traded in exchange for the credit that would keep Colonel Tom Parker, Presley's manager, at the tables, where he reportedly threw away millions of dollars.[83]

Yet Elvis's association with the city finds its truest expression in a song he recorded over a decade before he could have realized, in the mid-1970s, that there was no escape for him. 'Viva Las Vegas' (1964), an unmatched performance that inadvertently encapsulates the essence of a life given over to the pursuit of oblivion, identifies the city with a life of excess in its celebration of round-the-clock liberation, but also in figuring gambling as a restlessness that intensifies and accelerates the movement of a body into the kind of non-human transformation – to become the plaything of chance – that would be wrought in the ecstatic spaces of Las Vegas. The tempo of the performance matches the lyrical desire to keep raising the stakes and embracing the gods of chance one more time, regardless of the destruction of precious resources that would not be spent back in the real world of utility, an act of gratuitous waste. The performance of the song hones in on the fact that in Las Vegas money is transformed into something else – stripped of all its daily, taken-for-granted associations and 'metamorphosed', as Baudrillard wrote, 'into pure circulation, into pure fascination, into an absolute passion'.[84]

But what the song ultimately directs us to is the experience of the gaming room, in which, one might say, the world is turned inside out using means both subtle and nakedly obvious, should we care to look. Here, the regulated time of our everyday lives is made obsolete. The absence of clocks, along with the blocking-out of natural light (abolishing the division between day and night), is a renunciation of normal space-time conditions and expectations. Money, and the way it changes hands, its transformation from one kind of everyday value into something else, something more disposable, is what the city may ultimately be about. There is, on the one hand, the alleged dirty money that many would argue the city was established to make 'disappear' like nowhere else.[85] On the other hand, one cannot overlook the millions of dollars that are speedily lost and gained in the gaming rooms, something that remains mysterious but which formed a key scene in Martin Scorsese's 1995 film *Casino*, which peels back the curtain to reveal what goes on behind the scenes. In the counting rooms, money is moved on large trolleys and runs through anonymous hands and counting machines as a mob-appointed manager, Ace Rothstein (Robert De Niro), keeps an eye on it all. In voiceover, he tells the viewer that 'this is the end result of it all. The bright lights and broads and booze; it's all been arranged just to get your money.'[86]

Like the NTS, Las Vegas was run for decades in the belief that it could conceal its real workings under a cloak of forgetting. Rapt, the gamblers would temporarily lose sense of their own pasts and future, their sense of being, as they were caught in the ecstasy of the moment. For those who ran the place, however, its magical disappearing money, metaphorically sucked into the vast and capacious space of the desert, represented 'unearned, unaccounted for and untraceable money, forever'.[87] Or so it seemed. What cannot be denied, though, is that to achieve this forgetting took a wasteland setting – a desert – and the kind of disconnection that it could effect from time, memory and reality.

Conclusion: Data Wastelands

The contemporary technological sphere, particularly insofar as it allows us to consume virtual products and experiences in place of physical ones, dematerializes much of what once would have become waste. It perhaps leads us to assume that we might even make waste disappear. But hidden behind every virtual product or experience that seems no longer to need packaging or shipping – or to need the consumer to unpack a product and then dispose of it when it is no longer useful – are processes, server facilities and hidden physical infrastructures that nonetheless continue to consume resources and release pollutants, and thus contribute to the production of wastes, regardless of whether we see them or not. (That's not to mention the mining of rare minerals, which are required to power the digital and virtual world and its vast apparatuses.) This fact might bear out what I believe to be a truism: the apparent elimination of one kind of waste is only achieved at the expense of the creation of new waste processes. But there is more to it than that. In the contemporary situation, other phenomena related to the dematerialization of certain aspects of everyday life become apparent. You become yourself a source to be mined through your 24/7 connection to this virtual realm, in which 'the only consistent factor', according to critic Jonathan Crary, 'is the intensifying integration of one's time and activity into the parameters of electronic exchange'.[1] The 'option paralysis' common to life amid excessive virtual choice induces exhaustion and fatigue, qualities or conditions that are themselves the signs of minds or bodies that are being worn out as they become caught up in processes that they may have little awareness of.

But in this situation, all that changes is the modality of waste. Instead of the by-products of what you consume constituting waste, now your life – as expressed in your likes, your desires, your thoughts – becomes a new form of waste stuff; the overflow of your daily existence is now available as data and information that can be recycled and put to new uses by others. One aspect of this is evident in the way that the world is made available to be mined for data that we might personally find useful. Any phenomenon you choose to think of now seems to break down into a baffling array of classifications: new data points that are identified or tagged, for instance, by the way that they are labelled and the way in which these labels tend to multiply. This may reconfigure how we see the things of the world, and in doing so such data labels increase the quantities of invisible binary code that underlies the virtual realm and makes it serviceable to us. This is because living through computing technology has determined a reality in which everything now has to have as precise a definition as possible to be capable of being found within what cyberpunk author Bruce Sterling referred to, in 2010, as a 'colossally huge, searchable, public domain, which is now at your fingertips'.[2] The problem was that it could easily seem like one giant data wasteland, a resource that seemed to be 'all over the place, just termite mounds of poorly organized and extremely potent knowledge', all of it existing at the limit that separates us from it.[3] But it is in the more finely grained description of reality, and the ways that it can be brought into combinations or connect to systems and users that may exploit it further, that this seeming infinity of information that is held in data centres – the stuff of servers and retrieval systems that comes to life through our interactions – becomes valuable.

The detailed classification of popular music in this way provides an interesting example of the phenomenon, aiming as it does to make it easier to organize and serve up so-called content to consumers of digital platforms, such as Spotify, where access is presented to almost all recorded music that exists. The interaction of users with this vast and expanding archive illustrates the changes that this level of detail has brought to bear on how information is sorted and combined: new relations are generated between points of data and users of such data, who in turn generate their own data traces – or, we might say, digital waste trails – as a result of their own use and behaviour via the interactive technologies that make this content available.

Consider that, before the Internet existed, the musical styles or genres of previous eras didn't indicate that popular music was in any significant sense a fragmenting or collapsing form (although this might always have been implicit in its need or desire for reinvention). By the 1970s 'rock music' was differentiated from pop as a musical form because it placed greater emphasis on the performer as an autonomous artist driving stylistic change or innovations in the music as a form. Roxy Music was nothing like the Rolling Stones, but both groups were seen as examples of rock music. Today what were once stylistic differences are enough to give birth to a multitude of new 'genres'. According to Wikipedia, Roxy Music is said to belong to five sub-genres – 'art rock', 'glam rock', 'pop rock', 'art pop' and 'progressive rock'. Bryan Ferry – the writer of all the original Roxy Music songs, who also performed as a solo artist – additionally belongs to other genres, such as 'new wave' and 'sophisti-pop'. Ferry had always left himself space to develop in such a way that he was, as the performer, essentially continuous with the 'work of art' as the product that was consumed. But where he was able to lead a life that exemplified, in the words of one critic, 'the best possible example of the ultimate art-directed existence', the contemporary genres and labels that he is associated with are nothing to do with him.[4] They are the epiphenomenon of the information age, additional meaning overlaid on his body of work, reducing him to potentially new content to be further broken down and mined for unknown and possibly productive remaking. He will be recycled, brought into new combinations, and these labels represent the ways in which the logic of a digitally driven economy determines what Ferry is and how he may be described and consumed. This far exceeds the kind of 'packaging' or promotion that would have accompanied his earliest music at the time it was first released. And, in fact, everything that takes the form of information is susceptible to the same rearrangement, representation or, we might say, recycling.

The reformulation of written, recorded and other material online as 'content' in such ways signals a subtle change in how we think of concepts like information, material, documents, works of art, books, texts, opinions and even people. As *content* the immateriality of data that the proliferation of endless categories or data points represent points to a realm in which the stuff of information can be endlessly recycled and combined, made ready for new encounters that can fill up the time-space of online experience. This delivers, for example, the stuff that exists as diversions between advertisements, as we ceaselessly scroll or skim through what can be experienced as endlessly exhausting choices and options. The options come alive to engage the user and to feed back into invisible databases these interactions, which, once reduced to the condition of information, are easily amenable to algorithmic machine learning, with the aim to more efficiently analyse and monetize our attention the next time.

We exist on the cusp of information overflow, developing network dependence. We find the experience of the so-called attention economy in 'saturated' phenomena.[5] The immediacy of these momentary experiences manifests an oblivion that demands the 'ceaseless repetition' of the new mini-experience – TikTok pratfalls, YouTube cats (those most popular of unconscious reality performers in the present) and so on.[6] The complement to this type of experience is the unwelcome prospect of its negation in boredom, or what Harvie Ferguson has termed 'waste time'.[7] Boredom, Ferguson observes, 'is not to be confused with idleness, tedium, meaninglessness, or emptiness': 'It is not that "nothing happens" but that, in boredom, time stands aloof from all experiences that ordinarily appear effortlessly in flowing interconnection . . . it is time that cannot be "passed". Boredom is not wasted time, it is waste time.'[8] 'Waste time' is something projected onto a temporality that has not yet arrived but in which we become aware, so to speak, of a 'wasteland' within, eating us up for its own purpose, if not the glimpse of the self as a future ruin. For those whose aim is to turn our attention, and our submission to the saturated phenomena of the digital realm, into a new resource, the task is to find more efficient means of waste management. Our enjoyment is their resource. Thus, data must be subject to a range of operations that harness its nature as a thing with unseen potential. The data at some point may be harvested, reused, deleted, cleaned up and, of course, absorbed into cloud servers, thereafter relying on an invisible physical infrastructure of data centres whose energy consumption and polluting behaviour our digital lives feed, but which – as with other domains of waste – we remain largely oblivious to.[9]

Digital or virtual leftovers may have the character of material waste in the sense that they are difficult and time-consuming to manage, but, in truth, the condition of waste is a temporary one – everything becomes absorbed into, and sustains, a system that is global and interconnected. What there is, is an ontology of traces. In an essay from 1989, long before this was the case, titled 'Clues, Myths and the Historical Method', the historian Carlo Ginzburg said this: 'infinitesimal traces permit the comprehension of a deeper, otherwise unattainable reality.'[10] Waste becomes

Boredom and exhaustion meet – and are produced by – limitless 'content' in the digital age, as exemplified in the promise of escape through the interface of the always-present mobile connected device.

invisible and, in a manner of speaking, is absorbed into the 'atmosphere' that sustains life today. Early in the development of the microcomputer revolution, in the 1970s and '80s, it was thought that to realize the utopian promise of a networked society, and ultimately a world of connected users, information should be free. 'When you wrote a fine program you were building a community, not churning out a product,' notes Steven Levy in a history of the computer hackers who gave birth to the personal computer, first published in 1985.[11] In an afterword to a new edition of that book, Levy recalled that at a conference of hackers held in San Francisco the week his book was first published, Stewart Brand, founder of the *Whole Earth Catalog*, improvised his 'information should be free' idea in some off-the-cuff remarks during a session on what the future held in store for the hacker ethic of free access to everything that computers could make available. 'On the one hand information wants to be expensive, because it's so valuable,' Brand said, also noting: 'The right information in the right place just changes your life. On the other hand, information wants to be free, because the cost of getting it out is getting lower and lower all the time. So you have these two [tendencies] fighting against each other.'[12]

The data wastelands were opened up when the idea of services being provided for free became the basis for commercial activity to expand into the virtual realm. It has been a comforting modern myth that our wastes can be made to simply disappear without much friction. Like the stuff that we magically consign to oblivion by the action of a flush, or that we see vanish in the bin that is taken away by someone else, the material wastes that we generate have usually ended up in their own space: spirited away to the trash heaps and landfills that can be found on the peripheries of our cities. But just as, throughout history, the spaces people occupied and the habits they engaged in (from hunting and foraging for food to shopping in supermarkets) resulted in particular waste stuff that people were aware of because of its visibility or because of the smell of its lingering presence before it was disposed of, our own time and our current everyday practices give rise to their own peculiar waste.

In 2013 it was widely reported in the UK that a strange 'network' of recycling bins located in the City of London – the part of London where the financial services and banks are located – had been accumulating not only the usual trash but, unexpectedly, vast amounts of data from passing smartphone users. This was 'recycling' of a rather different kind.[13] These recycling bins, uniquely, were actually equipped with data-capturing technology. In the space of just one week, London's *Financial Times* reported, the smart devices of 4 million passers-by had valuable data siphoned off, which all disappeared invisibly into these bins (which is to say, through these disguised network interfaces). It seems that in the age of data exploitation, the immediate space that surrounds each of us becomes a data space, into which we leak information that may have a great commercial value to anyone interested in what we do, what we like or the kinds of people we know.[14] The fact that the revelations about the London 'spy bins' came at the same time as the revelations by hacker Edward Snowden, which exposed the extent of the data-gathering activities of national security agencies, may have been coincidental. But it raises the question of how the expansion of digital life within a new space of ever-expanding storage and digital memory that we come to rely on may leave us largely oblivious to the reality of this new immaterial waste that we are now producing as a matter of routine.

Every place, of course, will have known waste in the immense variety of its configurations during different periods of human life. People will also have known the hauntings or spectral intimations of waste – such as in the case of the City of London bins – when some entity makes off with waste that you didn't even know was a resource that you were producing for the benefit of others, who at the same time were concealing their nefarious

Data wastelands at Pionen Data Centre, from *TPB AFK: The Pirate Bay Away from Keyboard* (2013, dir. Simon Klose). The former underground defence facility in the Södermalm borough of Stockholm, Sweden, was converted for its current use by the Swedish Internet service provider Bahnhof in 2008.

data-gathering in the form of a litter bin. One has to marvel at the ingenuity of the mind that came up with the idea to use a structure so ubiquitous that it is invisible: swallowing up random bits and bytes that at some level of computational intelligence can be made sense of and reconciled into something meaningful, whose future value is as yet unknown. It is not unlike what Thomas Pynchon dreamt up in the paranoid 1960s when people in the underground counterculture sought to develop alternative postal systems to avoid governmental tampering with their mail. This idea animates his 1966 novel *The Crying of Lot 49*, where special mailboxes dotted around the urban landscape are, in fact, receptacles for the confused and meaningless scribblings of the conspiratorially minded and provide conduits to an underground communications network that goes by the acronym W.A.S.T.E.[15]

What we learn from the mutations that have produced transformations in the nature and idea of waste is that new forms of human life produce their own peculiar wastes. This is simply because to be human is to produce the world as an excess of meaning, or as something that leaves a remainder that is beyond meaning and comprehension, beyond the limits of our conscious engagement. No matter how we try to hide it.

REFERENCES

Introduction: Waste Is Life Plus Minus

1 Emanuel Perlmutter, 'Shots Are Fired in Refuse Strike, Filth Litters City', *New York Times*, 5 February 1968, p. 1.
2 Abbie Hoffman, *Revolution for the Hell of It* [1968] (New York, 2005), p. 55.
3 Ibid., p. 54.
4 Ibid., p. 55.
5 Jonah Raskin, *For the Hell of It: The Life and Times of Abbie Hoffman* (Berkeley, CA, 1996), p. 48.
6 Marty Jezer, *Abbie Hoffman, American Rebel* (New Brunswick, NJ, 1992), p. 128.
7 Quoted in Perlmutter, 'Shots Are Fired in Refuse Strike, Filth Litters City', p. 1.
8 Abbie Hoffman, *The Autobiography of Abbie Hoffman* (New York and London, 2000), p. 108.
9 Details of the film can be found in Jesse McKee, 'Because He Loved the People/Served the People? Counterculture Positions, Inheritance and Leadership', *C Magazine*, 126 (Summer 2015), pp. 18–22.
10 *Oxford English Dictionary*, 2nd edn, vol. XIX (Oxford, 1989), pp. 956–65.
11 William J. Mitchell, *Me++: The Cyborg Self and the Networked City* (Cambridge, MA, 2003).

1 Matter: Sewers, Filth and Sanitarians

1 Joseph C. Jenkins, *The Humanure Handbook: A Guide to Composting Human Manure* (Grove City, PA, 2005), p. 156.
2 Samuel Fishwick, 'Grease Is the Word', *Evening Standard*, 14 September 2017, p. 32.
3 Ibid., p. 33.
4 Matthew Gandy, 'The Paris Sewers and the Rationalization of Urban Space', *Transactions of the Institute of British Geographers*, XXIV/1 (1999), pp. 23–44 (p. 24).

5 See the account in Félix Nadar, *When I Was a Photographer*, trans. Eduardo Cadava and Liana Theodoratu (Cambridge, MA, 2015), pp. 75–94.
6 Walter Benjamin, *The Arcades Project*, trans. Howard Eiland and Kevin McLoughlin (Cambridge, MA, 1999), p. 6.
7 Nadar, *When I Was a Photographer*, p. 90.
8 John Hollingshead, *Underground London* (London, 1862), pp. 57–8.
9 'Rome's Cloaca Opened Again', *New York Times*, 3 November 1889, p. 17.
10 More info and background in John Hopkins, 'The "Sacred Sewer": Tradition and Religion in the Cloaca Maxima', in *Rome, Pollution and Propriety: Dirt, Disease and Hygiene in the Eternal City*, ed. Mark Bradley and Kenneth Stow (Cambridge, 2012), pp. 801–2.
11 Livy, *A History of Rome: Selections*, trans. and introduction Moses Hadas and Joe P. Poe (New York, 1962), section 56.
12 Pliny, *Natural History*, with English trans., 10 vols, ed. D. E. Eichholz, vol. X: *Libri XXXVI–XXXVII* [Loeb 419] (Cambridge, MA, and London, 1962), p. 83.
13 Ibid.
14 Ibid.
15 Louis Simond, *A Tour of Italy and Sicily* (London, 1828), p. 181.
16 Ibid.
17 Mary Beard, *SPQR: A History of Ancient Rome* (London, 2015), p. 25.
18 Nadar, *When I Was a Photographer*, p. 89.
19 Paul Ferris, 'Dugout Britain, or What Happens When the Bomb Drops', *The Observer*, 5 July 1981, pp. 25, 27.
20 'The Real Underworld', *Manchester Guardian*, 26 February 1925, p. 8.
21 Ibid.
22 Lewis Mumford, *Technics and Civilization* (New York, 1934).
23 Ibid., p. 357.
24 See Norbert Elias, *The Civilizing Process*, vol. I: *The History of Manners* (Oxford, 2000).
25 Robert Daley, *The World beneath the City* (Philadelphia, PA, and New York, 1959), p. 11.
26 Ibid.
27 Ibid., p. 12.
28 Ibid., p. 17.
29 Graham Harman, *Prince of Networks: Bruno Latour and Metaphysics* (Melbourne, 2009); Bruno Latour, *The Politics of Nature*, trans. C. Porter (Cambridge, MA, 2004), p. 237.
30 Kenneth J. Atchity, ed., *The Classical Roman Reader: New Encounters with Ancient Rome* (New York, 1997).
31 Michel Serres, *The Five Senses: A Philosophy of Mingled Bodies*, trans. Margaret Sankey and Peter Cowley (London and New York, 2008), p. 164.

32 Christopher Hamlin, 'Providence and Putrefaction: Victorian Sanitarians and the Natural Theology of Health and Disease', *Victorian Studies*, XXVIII/3 (Spring 1985), pp. 381–411 (pp. 381–2).
33 Mumford, *Technics and Civilization*, p. 170.
34 Peter Stallybrass and Allon White, *The Politics and Poetics of Transgression* (Ithaca, NY, 1986), p. 125.
35 Edwin Chadwick, *Report on the Sanitary Condition of the Labouring Classes of Britain* (London, 1842), p. 369.
36 Mary Poovey, *Making a Social Body: British Cultural Formation, 1830–1864* (Chicago, IL, and London, 1995), p. 40.
37 Deborah Cadbury, *Dreams of Iron and Steel* (New York, 2004), p. 121.
38 Ibid.
39 Lynda Nead, *Victorian Babylon: People, Streets and Images in Nineteenth-Century London* (New Haven, CT, 2000), p. 16.
40 Hamlin, 'Providence and Putrefaction', p. 383.
41 Ibid.
42 Ibid.
43 Ibid.
44 The letter is reproduced in John Meurig Thomas, *Michael Faraday and the Royal Institution* (Boca Raton, FL, 1991), pp. 132–3.
45 Thomas, *Michael Faraday and the Royal Institution*, pp. 132–3.
46 Charles Dickens, *Our Mutual Friend* (Oxford's World Classics) (Oxford, 2008), p. 1.
47 See, for example, Stephen Halliday, *The Great Stink of London: Sir Joseph Bazalgette and the Cleansing of the Victorian Metropolis* (Stroud, Gloucestershire, 2001).
48 Henry Mayhew, *London Labour and the London Poor*, vol. II (London, 1861), p. 386.
49 Robert Pogue Harrison, *The Dominion of the Dead* (Chicago, IL, 2003), p. 39.
50 L. Sprague de Camp and Catherine C. de Camp, *Ancient Ruins and Archaeology* (London, 1946), p. 85.
51 Ibid.
52 See Robert Garland, *Ancient Greece: Everyday Life in the Birthplace of Western Civilization* (New York, 2008), p. 141: 'In the opening scene of Aristophanes' "Women in Assembly", Blepyros defecates in the street as soon as he rises. Some houses, it seems, were not even provided with a cesspit (kapron). Women used a boat-shaped vessel called a skaphion. Although babies could be dangled out of the window in an emergency (see Aristophanes' "Clouds", line 1384), well-regulated houses possessed potties.'
53 Vito Fumagalli, *Landscapes of Fear: Perceptions of Nature and the City in the Middle Ages* (Cambridge, 1994).
54 Peter Ackroyd, *London: The Concise Biography* (London, 2012), p. 288.

55 Ibid.
56 Stephen Mennell, *All Manners of Food: Eating and Taste in England and France from the Middle Ages to the Present* (Urbana, IL, 1996), p. 310.
57 See Boria Sax, *City of Ravens: The Extraordinary History of London, the Tower and Its Famous Ravens* (London and New York, 2011), p. 47; the quote is from Ackroyd, *London*, p. 289.
58 Ernest L. Sabine, 'Butchering in Mediaeval London', *Speculum*, VIII/3 (1933), pp. 335–53; Ernest L. Sabine, 'Latrines and Cesspools of Medieval London', *Speculum*, IX/3 (1934), pp. 303–21; Ernest L. Sabine, 'City Cleaning in Mediaeval London', *Speculum*, XII/1 (1937), pp. 19–43.
59 Ackroyd, *London*, p. 290.
60 Ibid., p. 291.
61 Sabine, 'City Cleaning in Mediaeval London', p. 20.
62 Ibid.
63 Ibid.
64 Ibid., p. 21.
65 Ibid., p. 20.
66 Ibid., p. 21.
67 Ibid., p. 22.
68 Ibid.
69 Ibid.
70 Ibid.
71 The 'living bridge' of the Middle Ages is discussed in Ulf Strohmayer, 'Bridges: Different Conditions of Mobile Possibilities', in *Geographies of Mobilities: Practices, Spaces, Subjects*, ed. Tim Cresswell and Peter Merriman (Farnham, 2011), pp. 120–26, who notes that: 'In marked similarity to the late medieval urban fabric, "living bridges" were thus equipped with storage facilities, ateliers, shops, kitchens, latrines, sleeping rooms, as well as displaying a vertical social differentiation across the two or three floors initially constructed' (p. 124); quote from Lawrence Wright, *Clean and Decent: The Fascinating History of the Bathroom and the Water-Closet* (London, 1960), p. 50.
72 See Sabine, 'Butchering in Mediaeval London', p. 340.
73 Ibid., p. 343.
74 Ibid., p. 342.
75 Quoted in Kate Rigby, *Topographies of the Sacred: The Poetics of Place in European Romanticism* (Charlottesville, VA, 2004), p. 70.
76 David Frisby, *Cityscapes of Modernity: Critical Explorations* (Oxford, 2001), p. 55.
77 Ibid., p. 54.
78 Patrick Joyce, *The Rule of Freedom: Liberalism and the Modern City* (London, 2003), p. 14.
79 Ibid., p. 24.

80 I have explored this at greater length in John Scanlan, 'In Deadly Time: The "Lasting On" of Waste in Mayhew's London', *Time and Society*, XVI/2–3 (2007), pp. 189–206.
81 The quote is from the text by Simon Schaffer, 'Babbage's Intelligence', hosted at the Hypermedia Research Centre Archive, www.imaginaryfutures.net, accessed 11 January 2024. Note that this text differs in some respects from another published version: Simon Schaffer, 'Babbage's Intelligence: Calculating Engines and the Factory System', *Critical Inquiry*, XXI/1 (Autumn 1994), pp. 203–27. The latter is a shortened version of the former article and does not discuss Mayhew.
82 Mayhew, *London Labour and the London Poor*, vol. II, p. 155.
83 Ibid.
84 Catherine Gallagher, 'The Bio-Economics of "Our Mutual Friend"', in *Fragments for a History of the Human Body*, Part Three, ed. Michel Feher, Ramona Naddaff and Nadia Tazi (New York, 1989), pp. 344–65.
85 Mayhew, *London Labour and the London Poor*, vol. II, p. 155.
86 Ibid., p. 150.
87 Ibid., pp. 151–2.
88 Ibid., p. 150.
89 Ibid., pp. 151–2.
90 Ibid., p. 150.
91 Ibid., p. 124.
92 Ibid., p. 196.
93 Ibid.
94 Ibid.
95 Ibid.
96 Ibid.
97 Ibid., p. 9. The figures are quoted by Mayhew in pounds (*l*) and shillings (*s*).
98 Ibid., p. 7.
99 Ibid., p. 8.
100 Ibid.
101 See Schaffer, 'Babbage's Intelligence'.
102 Mayhew, *London Labour and the London Poor*, vol. II, p. 8.
103 On modernist ruins, see Owen Hatherley, *A Guide to the New Ruins of Great Britain* (London, 2011); images of the Garchey disposal system can be found in Peter Mitchell, *Memento Mori: The Flats at Quarry Hill, Leeds* (Bristol, 2016), p. 60.
104 The William Burroughs story was recounted in the BBC TV documentary *Arena: Burroughs* (1993, dir. Howard Brookner). The Bob Dylan X-ray hunters, traders in CT scans of Dylan's brain ('the ultimate bootleg'), are discussed in David Dalton, *Who Is That Man? In Search of the Real Bob Dylan* (Waterville, ME, 2012), p. 22. Mary Beard records that in ancient Rome, babies that looked 'weak or disabled' were often 'thrown away on

a rubbish tip'. See Beard, *SPQR*, p. 315. A campaign established in 1998 estimated that around three hundred Italian babies were thrown away, dumped in rubbish bins, every year. See Philip Willan, 'Italy's "Bin Babies" Find a Voice', *The Guardian*, 4 July 1998, p. 14.

2 Objects: Consume, Accumulate, Destroy

1 Peter Conrad, *Modern Times, Modern Places: Life and Art in the 20th Century* (London, 1998), p. 340.
2 Walter Benjamin, *The Arcades Project*, trans. Howard Eiland and Kevin McLaughlin (Cambridge, MA, 1999), p. 90.
3 Ibid., pp. 90–91.
4 Ibid.
5 Ibid.
6 Ben Highmore, *Everyday Life and Cultural Theory* (London and New York, 2002), p. 61.
7 Benjamin, *The Arcades Project*, p. 3.
8 Susan Buck-Morss, *The Dialectics of Seeing: Walter Benjamin and the Arcades Project* (Cambridge, MA, 1991), p. 65.
9 Benjamin, *The Arcades Project*, p. 540.
10 Ibid., p. 417.
11 David Frisby, *Cityscapes of Modernity: Critical Explorations* (Oxford, 2001), p. 30.
12 Benjamin, *The Arcades Project*, p. 417.
13 Robert Hughes, *The Shock of the New: Art and the Century of Change*, updated and enlarged edn (London, 1991).
14 Benjamin, *The Arcades Project*, p. 3.
15 Ibid.
16 Ibid., p. 4.
17 Ibid., p. 874.
18 Theodor W. Adorno, *Prisms*, trans. Shierry Weber Nicholsen and Samuel Weber (Cambridge, MA, 1983), p. 233.
19 Eric L. Santner, *On Creaturely Life: Rilke, Benjamin, Sebald* (Chicago, IL, 2006), p. 80.
20 Adorno, *Prisms*, p. 233.
21 Highmore, *Everyday Life and Cultural Theory*, pp. 60–74.
22 Benjamin, *The Arcades Project*, p. 14.
23 Esther Leslie, 'Walter Benjamin: Traces of Craft', *Journal of Design History*, XI/1 (1998), pp. 5–13 (p. 10).
24 Buck-Morss, *The Dialectics of Seeing*, p. 261.
25 Ibid., p. 268.
26 Frisby, *Cityscapes of Modernity*, p. 42.
27 Benjamin, *The Arcades Project*, p. 864.
28 Frisby, *Cityscapes of Modernity*, p. 41.

29 Richard Barbrook, *Imaginary Futures: From Thinking Machines to the Global Village* (London and Ann Arbor, MI, 2007), p. 27.
30 Daniel Boorstin, *The Decline of Radicalism: Reflections on America Today* (New York, 1963), p. 27.
31 See Barbrook, *Imaginary Futures*, p. 16.
32 Justin McGuirk, ed., *Waste Age: What Can Design Do?* (London, 2021), p. 46. For a perspective on the situation in Europe, see Heike Weber, 'Recycling, Remaking and Reusing in 20th-Century Europe: From a Culture of Thrift to One of Abundance', in *Throwaway: The History of a Modern Crisis*, ed. Christine Dupont, Stéphanie Gonçalves and Emma Teworte (Luxembourg, 2023), pp. 94–107.
33 Giles Slade, *Made to Break: Technology and Obsolescence in America* (Cambridge, MA, and London, 2006), p. 15.
34 Rachel Bowlby, *Carried Away: The Invention of Modern Shopping* (London, 2000).
35 Joseph A. Amato, *Dust: A History of the Small and the Invisible* (Berkeley, CA, 2000), pp. 80–81.
36 Vance Packard, *The Waste Makers* (Harmondsworth, 1963), p. 49.
37 Thomas Hine, *Populuxe* (Woodstock, NY, 2007), p. vi.
38 Packard, *The Waste Makers*, p. 50.
39 Ibid., p. 49.
40 J. Harry Dubois, *Plastics History USA* (Boston, MA, 1972), p. 1.
41 Ibid.
42 Ibid., p. 280.
43 Ibid., p. 365.
44 McGuirk, *Waste Age*, p. 54.
45 Gaston Bachelard, *The Poetics of Space*, trans. Maria Jolas, with a new foreword by John Stilgoe (Boston, MA, 1994), p. 47.
46 Mihaly Csikszentmihalyi and Eugene Rochberg-Halton, *The Meaning of Things: Domestic Symbols and the Self* (Cambridge, 1981), p. 184.
47 Andy Warhol, *The Philosophy of Andy Warhol* (New York, 1975), p. 144.
48 As described in Martin Melosi, *Garbage in the Cities: Refuse, Reform, and the Environment, 1880–1980* (College Station, TX, 1981), see pp. 42, 70, 113, 216 on late nineteenth-century and early twentieth-century practices. More recently, New Jersey as a dumping ground for New York is explored in Mira Engler, *Designing America's Waste Landscapes* (Baltimore, MD, 2004), pp. 76–7.
49 See, for example, John Urry, *Mobilities* (Cambridge, 2007).
50 Thomas Harris, *The Silence of the Lambs* (New York, 1988), p. 40.
51 Anne Caborn, 'A Neat and Secure Future's in Store', *The Observer*, 26 January 2003, p. 20.
52 See also my treatment in John Scanlan, *On Garbage* (London, 2005), pp. 173–8.

53 Jon Mooallem, 'The Self-Storage Self', *New York Times Magazine*, 6 September 2009, pp. 27–8.
54 John Steele, 'IRA Gang Grabs Bomb Cache', *Daily Telegraph*, 13 November 1996, pp. 1–2.
55 Mick Brown, 'Eno Puts Art in a Box', *Daily Telegraph*, 1 April 1995, p. 16.
56 The RCA students were referred to as the 'RCA Acorn Research Cell'. They were David Blame, Elaine Brechin, Michael Callan, Jason Coburn, Jason Edwards, Michelle Griffiths, Rachel Hale, Patricia Hepp, Tim Hutchinson, Chris Jones, Bettina Kubanek, Louise Lattimore, Danny Lavi, J. E. Lewis, Lucy McDonald, Natasha Michaels, Tim Noble, Hannah Redler, Janek Schaefer, Simon Waterfall and Sue Webster. See Artangel, *Off Limits: 40 Artangel Projects* (London and New York, 2002), p. 129.
57 Brown, 'Eno Puts Art in a Box', p. 16.
58 Kevin Jackson, 'Open Wide, This Won't Hurt', *The Independent*, 6 April 1995, p. 26.
59 Kevin Kelly, 'Eno: Gossip Is Philosophy', *Wired* (May 1995), p. 208.
60 Brian Eno, *A Year with Swollen Appendices: Brian Eno's Diary* (London, 1996), p. 72.
61 Ibid., p. 80.
62 Giles Coren, 'A Few Surprises in Store', *The Times*, 14 April 1995, p. 30.
63 Brown, 'Eno Puts Art in a Box', p. 16.
64 Laurie Anderson's voice, transcribed from a radio segment on the *Self Storage* installation, broadcast during an episode of BBC Radio 3's 'Mixing It', 3 April 1995.
65 James Hall, 'The White Stuff', *The Guardian* (G2 supplement), 25 April 1995, p. 5.
66 Coren, 'A Few Surprises in Store', p. 30.
67 Iain Sinclair, *Lights Out for the Territory: 9 Excursions in the Secret History of London* (London, 1997), p. 314.
68 Iain Sinclair, *Ghost Milk: Calling Time on the Grand Project* (London, 2011), p. 13.
69 This was in the section 'Business News', *Daily Telegraph*, 10 April 1990, p. 25.
70 Tim Dowling, 'Coming Soon', *Daily Telegraph* (magazine supplement), 31 December 1999, p. 8.
71 Iain Sinclair, *Hackney, That Rose-Red Empire: A Confidential Report* (London, 2009), p. 53.
72 Ibid.
73 Geoffrey Macnab, 'Portrait of the Artist as an Alluring Enigma', *The Independent*, 18 July 2014, p. 44.
74 James Cahill, 'Vandal in the Temple of High Art: Michael Landy', *Elephant: The Art and Visual Culture Magazine*, XV (2013), pp. 118–27 (p. 126).

75 Alex Coles, 'Site Specifics', in *Off Limits*, ed. Artangel, pp. 212–16 (p. 212).
76 Ibid.
77 See Michael Landy, 'Break Down', in *Off Limits*, ed. Artangel, pp. 162–7 (p. 162).
78 See, for instance, Dario Gamboni, *The Destruction of Art: Iconoclasm and Vandalism since the French Revolution* (London, 2007); Kerry Brougher, Russell Ferguson and Dario Gamboni, *Damage Control: Art and Destruction since 1950* (Munich, London and New York, 2013).
79 Michael Landy, 'Compulsory Obsolescence: A Series of Illustrations by Michael Landy Created during the Documentation Process Leading Up to *Break Down* (2001)', www.artangel.org.uk, accessed 7 April 2023.
80 Kerry Brougher, 'Radiation Made Visible', in *Damage Control*, ed. Brougher, Ferguson and Gamboni, p. 57.
81 Cahill, 'Vandal in the Temple of High Art', p. 126.
82 See, for example, the account of the importance of the Royal Academy of Arts' 1997 'Sensation' show in Gregor Muir, *Lucky Kunst: The Rise and Fall of Young British Art* (London, 2009), pp. 196–205.
83 This text is from an Artangel leaflet accompanying the event.
84 Derek Gale, 'Inspiration Born of Destruction' (letter to the editor), *Financial Times*, 22 February 2001, p. 18.
85 Muir, *Lucky Kunst*, p. 70.
86 Andrea Felsted and Daneshkhu Scheherazade, 'Louis Vuitton Opens Luxury Store', *Financial Times*, 27 May 2010, p. 20.
87 The reference is to Elizabeth Wilson, *Adorned in Dreams: Fashion and Modernity* (Berkeley, CA, 1987).
88 Matthew Collings, *Art Crazy Nation: The Post-Blimey! Art World* (London, 2001), p. 82.
89 Josh Spero, 'The Disassembly Line: How Michael Landy Tore His Life Apart', *Financial Times*, 30 January 2021.
90 Louisa Buck, 'Michael Landy's Epic Disassembly Line Revisited', *Art Newspaper*, 333 (April 2021), p. 50.
91 Jane Ure-Smith, 'Throwaway Culture: Michael Landy', *Financial Times* (Life and Arts supplement), 16 January 2010, p. 12.
92 Don DeLillo, *Underworld* (New York, 1997), p. 185.
93 Ibid., p. 102.
94 Ibid., p. 184.
95 Ibid., p. 286.
96 The quote is from the entry for 'Landfill' in Steve Coffel, *Encyclopedia of Garbage* (New York, 1996), p. 141.
97 Ibid.
98 Tim Dee, *Landfill: Gulls, Gull-Watchers and the Organising of Life* (Beaminster, Dorset, 2018), p. 15.
99 Ibid.

100 Mira Engler, *Designing America's Waste Landscapes* (Baltimore, MD, 2004), p. 95.
101 Gilles Lipovetsky, *The Empire of Fashion* (Princeton, NJ, 1994), p. 153.
102 Zygmunt Bauman, *Wasted Lives: Modernity and Its Outcasts* (Cambridge, 2003), pp. 63–93.
103 Peter Wynn Kirby, *Troubled Natures: Waste, Environment, Japan* (Honolulu, HI, 2011), p. 7.
104 Scanlan, *On Garbage*, pp. 22–33.
105 Sherry Turkle, ed., *Evocative Objects: Things We Think With* (Cambridge, MA, 2007).

3 Resources: Reclaim, Recover, Recycle

1 Lawrence R. Samuel, *Future: A Recent History* (Austin, TX, 2009), p. 2.
2 Simon Goldhill, *The Christian Invention of Time: Temporality and the Literature of Late Antiquity* (Cambridge, 2022), p. 86.
3 Ibid.
4 See Jacques Le Goff, *Medieval Civilization, 400–1500* (Oxford, 1984), p. 120.
5 Ibid.
6 Ibid.
7 Vito Fumagalli, *Landscapes of Fear: Perceptions of Nature and the City in the Middle Ages* (Cambridge, 1994), p. 72.
8 Le Goff, *Medieval Civilization*, p. 120.
9 G. Dohrn-van Rossum, *History of the Hour: Clocks and Modern Temporal Orders*, trans. Thomas Dunlap (Chicago, IL, 1996), p. 10.
10 Carlo M. Cipolla and Derek Birdsall, *The Technology of Man: A Visual History* (New York, 1979), p. 97.
11 Le Goff, *Medieval Civilization*, p. 177.
12 Ibid.
13 Lynn White Jr, 'The Historical Roots of Our Ecologic Crisis', *Science*, CLV/3767 (10 March 1967), pp. 1203–7 (p. 1205).
14 Eviatar Zerubavel, *Time Maps: Collective Memory and the Shape of the Past* (Chicago, IL, 2004), p. 17.
15 Svetlana Boym, *The Future of Nostalgia* (New York, 2001), p. 9.
16 Jacques Le Goff, *Time, Work, and Culture in the Middle Ages* (Chicago, IL, 1980), pp. 50–51.
17 Ibid.
18 Max Weber, *The Protestant Ethic and the Spirit of Capitalism*, trans. Talcott Parsons (New York and London, 1992), p. 157.
19 Max Weber, *General Economic History*, trans. Frank H. Knight (Glencoe, IL, 1950), p. 367.
20 Dohrn-van Rossum, *History of the Hour*, p. 10.
21 Matei Calinescu, *Five Faces of Modernity* (Durham, NC, 1987), p. 63.

22 W. G. Hoskins, *The Making of the English Landscape* (London, 1988), pp. 140–66; Michael Williams, 'The Enclosure and Reclamation of Waste Land in England and Wales in the Eighteenth and Nineteenth Centuries', *Transactions of the British Institute of Geographers*, 51 (1970), pp. 55–69.

23 Quoted in Clarence J. Glacken, *Traces on the Rhodian Shore: Nature and Culture in Western Thought from Ancient Times to the End of the Eighteenth Century* (Berkeley, CA, 1967), p. 153. Glacken is quoting from a commentary on the Book of Genesis: George E. Wright and Reginald H. Fuller, *The Book of the Acts of God* (Garden City, NY, 1960), p. 51.

24 Thomas More, *Utopia*, revd edn, ed. George M. Logan and Robert M. Adams (Cambridge, 2002), p. 54.

25 See Kate Rigby, *Topographies of the Sacred: The Poetics of Place in European Romanticism* (Charlottesville, VA, 2004), p. 64: 'In England the most significant vehicle of agricultural modernization was the parliamentary enclosure of the old collectively worked open fields and commons. Some areas in southern England had already been enclosed following the felling of the ancient forests in pre-Roman, Saxon, and medieval times. Elsewhere, particularly in the Midlands, large landowners had been enclosing smaller properties and common land since the fifteenth century in order to create more extensive sheep pastures. Until the mid-eighteenth century, however, roughly half the arable land in England, again mainly in the Midlands, plus large areas of "wild," "waste," and "common" land throughout the country, was still unenclosed.'

26 Joseph A. Amato, *Surfaces: A History* (Berkeley, CA, 2013), p. 140.

27 Gottfried Wilhelm Leibniz, 'Monadology' [§69], in *Philosophical Writings*, ed. G.H.R. Parkinson (London, 1973), p. 190.

28 Peter Sloterdijk, 'Lecture: Inspiration', transcribed and ed. Luc Peters, *Ephemera: Theory and Politics in Organization*, IX/3 (2009), pp. 242–51 (p. 245).

29 Eviatar Zerubavel, *Hidden Rhythms: Schedules and Calendars in Social Life* (Chicago, IL, 1981), p. 56.

30 Susan Buck-Morss, *Dreamworld and Catastrophe: The Passing of Mass Utopia in East and West* (Cambridge, MA, 2002), p. 103.

31 Ibid.

32 Martin Pawley, *Building for Tomorrow: Putting Waste to Work* (San Francisco, CA, 1982), p. 41.

33 Martin Pawley, *Garbage Housing* (London, 1975), p. 19.

34 Ibid., p. 47.

35 A. Seidenbaum, 'Students Build from Waste Materials', *Los Angeles Times* (section 7, Real Estate), 4 September 1977, p. 1.

36 See U.S. Housing Market Conditions, Historical Data, www.huduser.gov/periodicals/ushmc/winter2001/histdat08.htm, accessed 16 January 2024.

37 Seidenbaum, 'Students Build from Waste Materials', p. 6.
38 Yi-Fu Tuan, 'Environmental Attitudes', *Science Studies*, I/II (April 1971), pp. 215–24 (p. 215).
39 Rachel Carson, *Silent Spring* (New York, 1962).
40 Tuan, 'Environmental Attitudes', p. 215.
41 White, 'The Historical Roots of Our Ecologic Crisis', p. 1203; for a deeper exploration of the meanings, uses and implications of ecological systems, see Edward Goldsmith, *The Way: An Ecological World-View* (Athens, GA, 1988).
42 White, 'The Historical Roots of Our Ecologic Crisis', p. 1205.
43 'The National Economy Exhibition', *The Spectator*, 18 July 1916, p. 40.
44 Fredric Jameson, *Postmodernism; or, The Cultural Logic of Late Capitalism* (London, 1991), p. 160.
45 Ernest Callenbach, *Ecotopia: A Novel about Ecology, People and Politics in 1999* (London, 1978), p. 77.
46 Ibid.
47 'A Waste Exchange', Letter to the Editor (signed 'Pro Bono Publico'), *Manchester Guardian*, 8 March 1864, p. 6.
48 'Waste Exchange to Be Started', *The Guardian*, 31 October 1974, p. 18.
49 For historical studies of the early modern era, see for example Jon Stobart and Ilja Van Damme, eds, *Modernity and the Second-Hand Trade: European Consumption Cultures and Practices, 1700–1900* (Houndmills and New York, 2010); for a more contemporary consideration see Nicky Gregson and Louise Crewe, *Second-Hand Cultures* (Oxford and New York, 2003).
50 Mike Davis, *City of Quartz: Excavating the Future in Los Angeles* (New York, 1990), p. 381.
51 Martin O'Brien, *A Crisis of Waste? Understanding the Rubbish Society* (New York and London, 2012), p. 100.
52 See, for example, the accounts in Paul Farley and Michael Symmons Roberts, *Edgelands: Journeys into England's True Wilderness* (London, 2012), pp. 34–6.
53 Susan Strasser, *Waste and Want: A Social History of Trash* (New York, 2000), pp. 114–15.
54 'The Rag-and-Bone Trade', *Manchester Guardian*, 8 September 1924, p. 4.
55 Strasser, *Waste and Want*, pp. 114--15.
56 The account here summarized is in James Burke, *Connections* (Boston, MA, 1978), p. 100. On the uses of bones, see Strasser, *Waste and Want*, pp. 102–6.
57 John H. Munro, 'Textile Technology', in *The Dictionary of the Middle Ages*, ed. Joseph R. Strayer et al., vol. XI (New York, 1982), pp. 699–711.
58 Simon Schaffer, 'The Earth's Fertility as a Social Fact in Early Modern Britain', in *Nature and Society in Historical Context*, ed. Mikulas Teich, Roy Porter and Bo Gustafsson (Cambridge, 1997), pp. 124–47 (p. 126).

59 'A Boom in the Rag-and-Bone Business: Hawkers Chasing the Golden Fleece', *Manchester Guardian*, 21 March 1951, p. 3.
60 Allison Cooper, *Rationing* (London, 2004), pp. 20–21.
61 'A Boom in the Rag-and-Bone Business', p. 3.
62 Ibid.
63 'A Day in the Life of the Rag-and-Bone Man', *Manchester Guardian*, 2 June 1958, p. 5.
64 'A Boom in the Rag-and-Bone Business', p. 3.
65 James Collins, 'Rags to Riches', *The Guardian*, 20 February 1976, p. 8.
66 Jerry Hopkins, 'Beatle Loathers Return: Britain's Teddy Boys', *Rolling Stone*, 103 (2 March 1972), p. 14.
67 McLaren and Westwood's recycling of past fashions is explored at greater length in John Scanlan, *Sex Pistols: Poison in the Machine* (London, 2016), pp. 17–40, 69–106.
68 Emmanuelle Dirix, *Dressing the Decades: Twentieth-Century Vintage Style* (New Haven, CT, 2016), pp. 160–61.
69 Simon Reynolds, *Retromania: Pop Culture's Addiction to Its Own Past* (London, 2011), p. 186.
70 Sarah Mower, 'Retro Mania', *The Guardian* (8 January 1987), p. 10.
71 See, for example, Paul Gorman, *The Look: Adventures in Pop and Rock Fashion* (London, 2001). For an American perspective, see Jennifer Le Zotte, *From Goodwill to Grunge: A History of Secondhand Styles and Alternative Economies* (Chapel Hill, NC, 2017).
72 Elizabeth Wilson, *Adorned in Dreams: Fashion and Modernity* (Berkeley, CA, 1987), p. 193.
73 William Gibson, 'My Obsession', in *Everyday eBay: Culture, Collecting and Desire*, ed. Ken Hillis and Michael Petit with Nathan Scott Epley (New York and London, 2012), pp. 26–7.
74 Jean Baudrillard, *Simulacra and Simulation*, trans. Sheila Faria Glaser (Ann Arbor, MI, 1994).
75 Ibid., p. 13.
76 See, for example, Edward Humes, 'You Can't Recycle Garbage', *Sierra Magazine*, CIV/4 (July/August 2019), pp. 24–31, 42.
77 *The Sopranos*, season 1, episode 5, 'College' (originally broadcast 7 February 1999).
78 Jesse Kavadlo, 'Recycling Authority: Don DeLillo's Waste Management', *Critique*, XLII/4 (Summer 2001), pp. 384–401.
79 Don DeLillo, *Underworld* (New York, 1997), p. 279.
80 Kathlyn Gay, *Global Garbage: Exporting Trash and Toxic Waste* (New York, 1992), p. 32.
81 Ibid., p. 17.
82 Harold Crooks, *Giants of Garbage: The Rise of the Global Waste Industry and the Politics of Pollution Control* (Boston, MA, 1993), p. xi.
83 Ibid.

84 Ibid.
85 Humes, 'You Can't Recycle Garbage', p. 25.
86 Ibid., p. 27.
87 See 'What Is Sellafield?', https://nda.blog.gov.uk, accessed 21 January 2024.
88 Petra Tjitske Kalshoven, 'Surface-Making in Nuclear Decommissioning: A Narrative of Sludge, Plutonium and Their Whereabouts', in *Surface and Apparition: The Immateriality of Modern Surface*, ed. Yeseung Lee (London, 2020), pp. 37–50 (p. 39).
89 Nuclear Waste Services, 'GDF Report 2023: Delivering a Solution for the Most Radioactive Waste' (2023), available at www.gov.uk, accessed 21 January 2024.

4 Aesthetics: Designing and Dematerializing

1 Reyner P. Banham, 'A Throw-Away Aesthetic', in *Design by Choice*, ed. Penny Sparke (New York, 1981), p. 90.
2 Ibid.
3 Stuart Ewen, *All Consuming Images: The Politics of Style in Contemporary Culture* (New York, 2002), p. 245.
4 *The Fall and Rise of Reginald Perrin*, based on a series of novels by David Nobbs, BBC1, 1977. The story of Grot is the basis of the second TV series.
5 Herbert Marcuse, *One-Dimensional Man: Studies in the Ideology of Advanced Industrial Society* (Boston, MA, 1964), p. 7.
6 Ellen Lupton and J. Abbott Miller, 'Hygiene, Cuisine and the Product World of Early Twentieth-Century America', in *Zone 6: Incorporations*, ed. Jonathan Crary and Sanford Kwinter (New York, 1992), p. 503.
7 Ibid.
8 Jason K. Miller, *Frank Gehry* (New York, 2002), p. 14.
9 Frank O. Gehry, *The Architecture of Frank Gehry* (New York, 1986), p. 63.
10 Kevin Starr, *Material Dreams: Southern California through the 1920s* (New York, 1990), p. 229.
11 Fredric Jameson, *Postmodernism; or, The Cultural Logic of Late Capitalism* (London, 1991), p. 111.
12 Aaron Betsky, 'Love It or Hate It, Gehry House Shows an Original Mind at Work', *Los Angeles Times*, 14 February 1991, p. J12.
13 Jameson, *Postmodernism*, p. 11.
14 Mike Davis, *City of Quartz: Excavating the Future in Los Angeles* (New York, 1990), p. 238.
15 John Dreyfus, 'Gehry's Artful House Offends, Baffles, Angers His Neighbors', *Los Angeles Times* (section 8, Real Estate), 23 July 1978, pp. 1, 24.
16 Ibid.

17 Ibid., p. 24.
18 Roseanne Robertson, 'Frank Gehry: Small Buildings Getting Better and Better', *Sydney Morning Herald*, 23 June 1982, p. 8.
19 Frank O. Gehry, 'Beyond Function', *House and Home: Design Quarterly*, 138 (1987), pp. 2–11 (p. 5).
20 Kevin Starr, *Coast of Dreams: A History of Contemporary California* (London, 2005), p. 57.
21 Martin Pawley, *20th Century Architecture: A Reader's Guide* (Oxford, 2000), p. 78.
22 Martin Melosi, *Garbage in the Cities: Refuse, Reform, and the Environment, 1880–1980* (College Station, TX, 1981), p. 168.
23 Ibid.
24 Ibid.
25 'New Resort's "Mountain" Will Be Built of Garbage', *Wichita Beacon*, 23 September 1967, p. 4.
26 Ibid.
27 John Lonsdale, 'Tyneside [Inertia]: Shifting Margins', in *Mutations*, ed. Rem Koolhaas, Harvard Project on the City et al. (Barcelona, 2000), p. 393.
28 Stefano Boeri, 'Notes for a Research Program', in *Mutations*, ed. Koolhaas et al., p. 375.
29 Iain Sinclair, 'A London View', in *Granta 65: London – Lives of the City* (London, 1999), p. 126.
30 Ibid.
31 See, for example, Thomas Hine, *The Total Package: The Evolution and Secret Meanings of Boxes, Bottles, Cans, and Tubes* (Boston, MA, 1995).
32 John Elkington and Julia Hailes, *The Green Consumer Guide: From Shampoo to Champagne, High Street Shopping for a Better Environment* (London, 1988); John Elkington and Tom Burke, *The Green Capitalists: Making Business Sense of the Environment* (London, 1987).
33 Roger Highfield, 'Caught in the Aerosol Trap', *Daily Telegraph*, 9 January 1989, p. 15.
34 Charles Glover, 'Red Alert in the Green Year', *Daily Telegraph* (Weekend section), 31 December 1988, p. 10.
35 Ibid.
36 Ibid.
37 Don DeLillo, *White Noise* (New York, 1985).
38 Gordon Burn, 'Wired Up and Whacked Out', *Sunday Times Magazine*, 25 August 1991, p. 36.
39 Elkington and Burke, *The Green Capitalists*, p. 11.
40 Paul Hoyland, 'Speculation on Leadership Is Scorned by Icke', *The Guardian*, 22 September 1989, p. 6.
41 Ibid.
42 Robin Murray, *Zero Waste* (London, 2002), p. 3. Murray explains that 'The term Zero Waste has its origins in the highly successful Japanese

industrial concept of total quality management (TQM). It is influenced by ideas such as "zero defects", the extraordinarily successful approach whereby producers like Toshiba have achieved results as low as one defect per million. Transferred to the arena of municipal waste, Zero Waste forces attention onto the whole lifecycle of products.'

43 Ibid.

44 Paul Judge et al., 'Progress Unlocked', *RSA Journal*, CL/5509 (October 2003), pp. 18–25.

45 Pauline Madge, 'Ecological Design: A New Critique', *Design Issues*, XIII/2 (Summer 1997), pp. 44–54 (p. 51).

46 Justin McGuirk, ed., *Waste Age: What Can Design Do?* (London, 2021), p. 204.

47 Most of these examples are from issues of the technology magazine *Wired*. See also McGuirk, *Waste Age*.

48 Anna Maria Church, 'Martí Guixé: A New Point of View', *T Magazine*, 23 January 2019, available at www.tmagazine.es.

49 Jonathan Crary, *24/7: Late Capitalism and the Ends of Sleep* (London, 2013), p. 40.

50 See 'Theory of the Quasi-Object', in Michel Serres, *The Parasite*, trans. Lawrence R. Scher (Baltimore, MD, and London, 2007), pp. 224–34.

51 Sherry Turkle, ed., *Evocative Objects: Things We Think With* (Cambridge, MA, 2007).

52 Ibid., p. 8.

53 McGuirk, *Waste Age*, p. 90.

54 Ibid., p. 250.

55 Karen Ghung, 'The Age of Techneco', *Wallpaper*, 92 (2006), pp. 135–40.

56 Martin Pawley, *Garbage Housing* (London, 1975).

57 Lynn Spigel, 'Designing the Smarthouse: Posthuman Domesticity and Conspicuous Production', in *Electronic Elsewheres: Media, Technology, and the Experience of Social Space*, ed. Chris Berry, Soyoung Kim and Lynn Spigel (Minneapolis, MN, 2010), pp. 59–92 (p. 62).

58 Charles Bevier, 'Industry Brief', *Building Systems Magazine* (March 2000), p. 10.

59 Alvin Toffler, *Future Shock* (New York, 1970).

60 Ibid., p. 6.

61 Joseph Cuomo, 'A Conversation with W. G. Sebald', in *The Emergence of Memory: Conversations with W. G. Sebald*, ed. Lynn Sharon Schwartz (New York, 2007), p. 114.

62 Don DeLillo, *Libra*, with a new introduction by the author (London, 2006).

63 Adam Begly, 'Don DeLillo: The Art of Fiction No. 135', *Paris Review*, 128 (Fall 1993), available at www.theparisreview.org.

64 DeLillo, 'Assassination Aura', new introduction to *Libra*, p. ix.

65 Susan Sontag, *On Photography* (London, 1978), p. 179.

66 Ibid.
67 Ibid.
68 Burn, 'Wired Up and Whacked Out', p. 36.
69 Patrick Stokes, *Digital Souls: A Philosophy of Online Death* (London, 2021), p. 95.
70 See Brian Hayes, 'Bit Rot', *American Scientist*, LXXXVI/5 (September–October 1998), pp. 410–15.
71 Ibid., p. 412.
72 Victoria Sloyan, 'Born-Digital Archives at the Wellcome Library: Appraisal and Sensitivity Review of Two Hard Drives', *Archives and Records*, XXXVII/1 (2016), pp. 20–36 (p. 22).
73 Ibid.
74 Marshall McLuhan and Quentin Fiore, *The Medium Is the Massage*, coordinated by Jerome Agel (New York, 1967), p. 68.
75 Ibid., p. 63.
76 Peter N. Limberg and Conor Barnes, 'The Memetic Tribes of Culture War 2.0', *Medium*, www.medium.com, accessed 26 May 2023.
77 McLuhan and Fiore, *The Medium Is the Massage*, p. 68.
78 Geoff Cox, 'Generator: The Value of Software Art', in *Issues in Curating Contemporary Art and Performance*, ed. Judith Rigg and Michèle Sedgwick (Bristol, 2007), pp. 147–62 (p. 160).
79 Stuart Brisley, *Beyond Reason: Ordure* (London, 2003).
80 Ibid., section 5.
81 Caleb Kelly, *Cracked Media: The Sound of Malfunction* (Cambridge, MA, 2009), p. 4.
82 Jim McLellan, 'A Scanner in the Works', *The Observer* (Life section), 26 March 1995, p. 66.

5 Projections: Wastelands, Real and Imagined

1 Mary Shelley, *The Last Man*, ed. with introduction and notes by Morton D. Paley (Oxford and New York, 1998), pp. 469–70.
2 H. G. Wells, *The Time Machine* (abridged) (London, 1996), p. 20.
3 Ibid., p. 17.
4 H. G. Wells, *War of the Worlds* [1898] (New York, 1986), p. 50.
5 John Carey, *The Intellectuals and the Masses: Pride and Prejudice among the Literary Intelligentsia, 1880–1939* (London, 1992), p. 130.
6 Ibid., pp. 3–22.
7 Tim Edensor, 'The Ghosts of Industrial Ruins: Ordering and Disordering Memory in Excessive Space', *Environment and Planning D: Society and Space*, XXIII/6 (2005), pp. 829–49.
8 See, for example, Camilo José Vergara, *American Ruins* (New York, 1999).
9 Boris Sagal (dir.), *The Omega Man* (Walter Seltzer Productions, 1971).

10 Thom Andersen (dir.), *Los Angeles Plays Itself* (Cinema Guild, 2013). It should be noted that this film, originally released in 2003, aside from widely circulated bootleg digital versions, was only played at film festivals and, as such, had no distributor and no official release. The 2013 DVD release is the only version that exists as a point of reference.
11 Quote from David Frum, *How We Got Here, the 70s: The Decade That Brought You Modern Life* (New York, 2000), p. xxiii.
12 As discussed in John Scanlan, *Van Halen: Exuberant California, Zen Rock'n'roll* (London, 2012), pp. 26–7.
13 Mike Davis, *City of Quartz: Excavating the Future in Los Angeles* (London and New York, 1990), p. 18.
14 Ibid., p. 37.
15 For example, 'Far from gloating about the fictional destruction of Los Angeles in books and films, I argue that this has become an almost fascist phenomenon,' in Joshuah Bearman, 'An Interview with Mike Davis', *The Believer*, 10 (February 2004), available at www.thebeliever.net.
16 Mike Davis, *Ecology of Fear: Los Angeles and the Imagination of Disaster* (London, 1999), p. 276.
17 Ibid.
18 Ibid., p. 278.
19 Jean Baudrillard, *America*, trans. Chris Turner (London, 1988), pp. 51–2.
20 Bart Mills, 'Marlowe Meets *Alien*', *The Guardian*, 15 June 1981, p. 9.
21 Ibid.
22 Philip K. Dick, *Do Androids Dream of Electric Sheep?* [1968] (Waterville, ME, 2001), p. 76.
23 Ridley Scott (dir.), *Blade Runner* (Blade Runner Partnership, The Ladd Company, Run Run Shaw, Warner Bros., 1982).
24 Dick, *Do Androids Dream of Electric Sheep?*, p. 27.
25 Davis, *Ecology of Fear*, p. 337.
26 Quoted in Kerwin Lee Klein, 'Westward, Utopia: Robert V. Hine, Aldous Huxley, and the Future of California History', in *Bloom's Modern Critical Views: Aldous Huxley*, ed. with an introduction by Harold Bloom (Philadelphia, PA, 2003), p. 154.
27 Ibid., p. 155.
28 Ibid.
29 Dominic Priore, 'The Decline of Western Civilization', in *The Decline of Western Civilization Collection*, DVD booklet essay (Los Angeles, CA, 2015), p. 5.
30 Harvie Ferguson, 'Exteriority: Boredom, Disgust, and the Margins of Humanity', in *Aesthetic Fatigue: Modernity and the Language of Waste*, ed. John Scanlan and John F. M. Clark (Newcastle upon Tyne, 2013), pp. 220–50 (p. 222).
31 Dave Thompson, *London's Burning: True Adventures in the Frontline of Punk, 1976–1977* (Chicago, IL, 2009), p. 161.

32 Steve Turner, 'The Anarchic Rock of the Young and Doleful', *The Guardian*, 3 December 1976, p. 13.
33 Clive James, 'A Load of Punk', *The Observer*, 5 December 1976, p. 29.
34 James Wolcott, 'Kiss Me, You Fool: Sex Pistols '77', *Village Voice*, 21 November 1977, p. 53.
35 Iain Gray, 'The Punk World: Swastikas, Rubbish Bags, Torn Clothes', *Glasgow Herald*, 22 August 1977, p. 7.
36 Greil Marcus, *Lipstick Traces: A Secret History of the Twentieth Century* (Cambridge, MA, 1989), p. 27.
37 The Sadia Ltd advertisement referred to here appears in *The Municipal Journal*, 29 September 1961, p. 3095.
38 Owen Hatherley, *A Guide to the New Ruins of Great Britain* (London, 2010), p. 118.
39 Ivan Chtcheglov, 'Formulary for a New Urbanism', in *Situationist International Anthology*, ed. Ken Knabb (Berkeley, CA, 1989), pp. 2–3.
40 Martin Amis, *The War against Cliché: Essays and Reviews, 1971–2000* (New York, 2001), p. 116.
41 J. G. Ballard, *J. G. Ballard: Quotes*, ed. V. Vale and Mike Ryan (San Francisco, CA, 2004), p. 158.
42 Rick Poyner, *Oh So Pretty: Punk in Print, 1976–80* (London and New York, 2016), p. 21.
43 William Gibson, '1977', in *Punk: An Aesthetic*, ed. Johan Kugelberg and Jon Savage (New York, 2012), p. 286.
44 Dick Hebdige, *Subculture: The Meaning of Style* (London and New York, 1998), p. 107.
45 Legs McNeil and Gillian McCain, *Please Kill Me: The Uncensored Oral History of Punk* (New York, 1997), p. 208.
46 Jon Savage, *The England's Dreaming Tapes* (London, 2009), p. 390.
47 Ibid., p. 284.
48 Vivienne Westwood and Ian Kelly, *Vivienne Westwood* (London, 2014), p. 199.
49 Angela Neustatter, 'Punktuation', *The Guardian*, 23 September 1977, p. 11.
50 Johan Kugelberg, 'The Last Macro Tribe: In Conversation with William Gibson and Jon Savage', in *Punk: An Aesthetic*, ed. Kugelberg and Savage, p. 346.
51 Savage, *England's Dreaming*, p. 284.
52 Michael Bracewell, 'Punk', in *London: From Punk to Blair*, ed. Joe Kerr and Andrew Gibson (London, 2003), pp. 301–6 (p. 303).
53 Neustatter, 'Punktuation', p. 11.
54 Westwood and Kelly, *Vivienne Westwood*, p. 158.
55 Vagrancy Act 1824, Section 4, 'Persons committing certain offences to be deemed rogues and vagabonds', see www.legislation.gov.uk/ukpga/Geo4/5/83, accessed 17 September 2024.

56 See the account in John Scanlan, *Sex Pistols: Poison in the Machine* (London, 2016), pp. 205–10.
57 Neustatter, 'Punktuation', p. 11.
58 Jonathan Ross, *The Incredibly Strange Film Show: Russ Meyer*, Channel 4 TV (UK), 9 September 1988.
59 Peter Ackroyd, *London: The Concise Biography* (London, 2012), p. 616.
60 Clare Dover, 'The Barefoot Trail to Rotherhithe', *Daily Telegraph*, 1 June 1976, p. 8.
61 Ibid.
62 Fiona Cantell, 'Rotherhithe Street Festival', *The Ecologist*, VI/4 (1976), p. 146.
63 Pamela Church Gibson, 'Imaginary Landscapes, Jumbled Topographies: Cinematic London', in *London: From Punk to Blair*, ed. Kerr and Gibson, pp. 363–70 (p. 364).
64 Savage, *England's Dreaming*, p. 666.
65 Rachel Barnes, *The Pre-Raphaelites and Their World* (London, 1998), p. 85.
66 For a fuller account of the making of the film, see Michael O'Pray, *Derek Jarman: Dreams of England* (London, 1996), pp. 153–66.
67 See the descriptions of Dungeness in Michael Charlesworth, *Derek Jarman* (London, 2011), pp. 11–18.
68 William Raban (dir.), *William Raban: British Artists' Films* (BFI, 2004).
69 Jonathan Davis, 'Docklands – 5,000 Acres of New Hope', *Sunday Telegraph*, 8 November 1981, p. 26.
70 The advert appeared in the *Daily Telegraph*, 25 April 1988, p. 3.
71 Charles Nevin, 'A Firm Step into the Wasteland', *Daily Telegraph*, 17 September 1987, p. 13.
72 Barnaby Wright, 'Creative Destruction: Frank Auerbach and the Rebuilding of London', in *Frank Auerbach: London Building Sites, 1952–1962*, exh. cat., Courtauld Gallery (London, 2009), pp. 12–35 (p. 14).
73 Barry Miles, *London Calling: A Countercultural History of London since 1945* (London, 2010), p. 97.
74 Elena Crippa, 'The Human Condition: Six Modern Painters Reinvent Reality', in *London Calling: Bacon, Freud, Kossoff, Andrews, Auerbach, and Kitaj*, ed. Elena Crippa and Catherine Lampert (Los Angeles, CA, and London, 2016), pp. 1–12 (p. 8).
75 See *Frank Auerbach: London Building Sites, 1952–1962*.
76 Adrian Heathfield, ed., *Live: Art and Performance* (New York, 2004), p. 8.
77 Adrian Henri, *Total Art: Environments, Happenings, and Performance* (New York, 1974), p. 128.
78 Melanie Roberts, 'Stuart Brisley', *National Life Stories: Artists' Lives* (London, 1996), https://sounds.bl.uk/Arts-literature-and-performance/Art, accessed 15 August 2022, sound recording.

79 Peter Davey and Kurt W. Forster, *Exploring Boundaries: The Architecture of Wilkinson Eyre* (Basel, 2007), p. 127.
80 The villains plotting a robbery in *The Ladykillers* are seen regularly disposing of people who are in the way by dumping them from a bridge onto coal trains passing below, which are heading to and from the marshalling yards of King's Cross station. In Joel and Ethan Coen's 2004 remake, set in the American South, the gang dumps dead bodies from a bridge onto garbage barges, which pass by on a canal below.
81 John R. Kellett, *The Impact of Railways on Victorian Cities* (London and Toronto, 1969), p. 247.
82 Henry Mayhew, *London Labour and the London Poor*, vol. IV (London, 1864), p. 309.
83 John Hollingshead, *Underground London* (London, 1862), p. 230.
84 Henry Mayhew, *London Labour and the London Poor*, vol. II (London, 1861), p. 406.
85 Stuart Brisley and Gilane Tawadros, *The Stuart Brisley Interviews: Performance and Its Afterlives* (London, 2020), p. 95.
86 Ibid., p. 96.
87 Richard Cork, *Everything Seemed Possible: Art in the 1970s* (New Haven, CT, and London, 2003), p. 181.
88 See Stuart Brisley, *Georgiana Collection*, exh. cat., Third Eye Centre, Glasgow, and others (Glasgow and Derry, 1986). The work referenced here was originally titled *Leaching out at and from the Intersection* (pp. 5–7 of this exhibition catalogue), but more recently the artist's website refers to it as *Leaching Out at the Intersection.*
89 Michael Archer, 'Neither One Thing nor the Other', in Brisley, *Georgiana Collection*, pp. 5–26 (p. 6).
90 William Feaver, 'Mr Brisley's Rubbish', *The Observer*, 10 May 1981, p. 35.
91 Ibid.
92 Roberts, 'Stuart Brisley'.
93 Stuart Brisley, 'Actions and Installations', *CV/Visual Arts Research*, III/1 (February 2014), p. 26.
94 Waldemar Januszczak, 'Breaking Down the Sound Barrier', *The Guardian*, 1 September 1982, p. 11.
95 Jane England, 'Introduction', in Brisley, *Stuart Brisley: Works, 1958–2006* (London, 2006), p. 5.
96 Hugh Herbert, 'Easy to Mock, Impossible to Explain . . . Stuart Brisley', *The Guardian*, 18 September 1978, p. 9.
97 Brisley, *Georgiana Collection*, p. 17.
98 Paul Gorman, *The Life and Times of Malcolm McLaren* (London, 2020), p. 497.
99 Jane Mulvagh, *Vivienne Westwood* (London, 2003), pp. 161–2.
100 Gorman, *The Life and Times of Malcolm McLaren*, p. 497.

101 Stuart Brisley, *Beyond Reason: Ordure* (London, 2003), section 4.
102 Ibid., section 47.

6 Temporalities: Deep, Infinite and Meaningless

1 *Early Greek Philosophy*, trans., ed. and with an introduction by Jonathan Barnes (London, 1987), p. 56.
2 W.K.C. Guthrie, *The Greek Philosophers: From Thales to Aristotle* (London, 1967), p. 27.
3 R. G. Collingwood, *The Idea of Nature* (New York, 1960), p. 33.
4 Guthrie, *The Greek Philosophers*, pp. 31–2.
5 George L. Thompson, 'Death Valley Visited: Trip of 320 Miles from Los Angeles Leads Motorists to Interesting Desert Land', *Los Angeles Times* (Automobile and Aviation section), 20 January 1929, p. 1.
6 John C. Van Tramp, *Prairie and Rocky Mountain Adventures; or, Life in the West* (Columbus, OH, 1860).
7 Christopher Reed, 'Find Reveals Death Valley's Dark Secrets', *The Guardian*, 21 January 1999, p. 13.
8 Thompson, 'Death Valley Visited', p. 1.
9 Hunter S. Thompson, *Fear and Loathing in Las Vegas* (New York, 1989), p. 3.
10 David Jasper, *Sacred Desert: Religion, Literature, Art and Culture* (Oxford, 2004), p. 109.
11 Edwin Bryant, *What I Saw in California* (Lincoln, NE, 1985), p. 135.
12 Jack Hicks et al., *The Literature of California*, vol. I: *Native American Beginnings to 1945* (Berkeley, CA, 2000), p. 129.
13 Jasper, *Sacred Desert*, p. 7.
14 Francesco Orlando, *Obsolete Objects in the Literary Imagination: Ruins, Relics, Rarities, Rubbish, Uninhabited Places, and Hidden Treasures*, trans. Gabriel Pihas, Daniel Seidel and Alessandra Grego (New Haven, CT, and London, 2006), p. 141.
15 Richard B. Onians, *The Origin of European Thought about the Body, the Soul, the Mind, the World, Time and Fate* (Cambridge, 1988), p. 221.
16 Martha C. Nussbaum, *The Fragility of Goodness: Luck and Ethics in Greek Tragedy and Philosophy* (Cambridge, 1986), p. 2.
17 Quoted in Clarence J. Glacken, *Traces on the Rhodian Shore: Nature and Culture in Western Thought from Ancient Times to the End of the Eighteenth Century* (Berkeley, CA, 1967), p. 14.
18 Reyner P. Banham, *Scenes in America Deserta* (London, 1982), p. 112.
19 Daniel Heller-Roazen, *Echolalias: On the Forgetting of Language* (New York, 2005), p. 53.
20 David Thomson, *In Nevada: The Land, the People, God and Chance* (London, 1999), p. 23.
21 Greg Bishop et al., *Weird California* (New York and London, 2006), p. 54.

22 Banham, *Scenes in America Deserta*, p. 37.
23 Len Wilcox, *Desert Dancing: Exploring the Land, the People, the Legends of the California Desert* (Walpole, MA, 2000), pp. 70–71.
24 Charles Hillinger, '"Squatter" Owes Rent, U.S. Says', *Los Angeles Times* (Part 2, Editorials), 14 July 1967, pp. 1, 6.
25 Ibid., p. 1.
26 See, for example, Kevin Starr, 'Carey McWilliams's California: The Light and the Dark', in *Reading California: Art, Image and Identity, 1900–2000*, ed. Stephanie Barron (Berkeley, CA, 2001), pp. 15–30; Erik Davis, *The Visionary State: A Journey through California's Spiritual Landscape* (San Francisco, CA, 2006).
27 Kevin Starr, *Coast of Dreams: A History of Contemporary California* (London, 2005), pp. 10–13.
28 Banham, *Scenes in America Deserta*, p. 63.
29 William L. Fox, *The Void, the Grid and the Sign: Traversing the Great Basin* (Reno, NV, 2005), p. 2.
30 Philip Sheldrake, *Spaces for the Sacred: Place, Memory, and Identity* (Baltimore, MD, 2001), p. 91.
31 John McPhee, *Basin and Range* (New York, 1981), pp. 220–29.
32 Ibid., p. 229.
33 Susan Danly, 'Charis Wilson and Edward Weston: California and the West', in *Perpetual Mirage: Photographic Narratives of the Desert West*, ed. May Castleberry (New York, 1996), pp. 125–6.
34 Nancy Newhall, *The Photographs of Edward Weston* (New York, 1946).
35 M. Kat Anderson, Michael G. Barbour and Valerie Whitworth, 'A World of Balance and Plenty: Land, Plants, Animals, and Humans in a Pre-European California', in *Contested Eden: California before the Gold Rush*, ed. Ramón Gutiérrez and Richard J. Orsi (Berkeley, CA, 1998), pp. 12–47.
36 Aldous Huxley, 'The Desert', in *West of the West: Imagining California, an Anthology*, ed. Leonard Michaels, David Reid and Raquel L. Scherr (San Francisco, CA, 1989), p. 322.
37 Banham, *Scenes in America Deserta*, p. 205.
38 Tom Vanderbilt, *Survival City: Adventures among the Ruins of Atomic America* (New York, 2002), p. 27.
39 John W. Day and Charles Hall, *America's Most Sustainable Cities and Regions: Surviving the 21st Century Megatrends* (New York, 2016), p. 94.
40 Sally Denton and Roger Morris, *The Money and the Power: The Making of Las Vegas and Its Hold on America, 1947–2000* (London, 2002), p. 98.
41 Banham, *Scenes in America Deserta*, p. 145.
42 Valerie L. Kuletz, *The Tainted Desert: Environmental and Social Ruin in the American West* (Abingdon and New York, 1998), p. 66.
43 The quote is by *New York Post* columnist Harriet van Horne, reproduced in an advert for the film that was published in the *New York Times*, 22 February 1970, p. D14.

44 Beverley Walker, 'Michelangelo and the Leviathan: The Making of Zabriskie Point', *Film Comment* (September 1992), pp. 36–49 (p. 38).
45 Ibid.
46 Ibid., p. 44.
47 James Riley, *The Bad Trip: Dark Omens, New Worlds and the End of the Sixties* (London, 2019), p. 98.
48 Ibid.
49 Vincent Bugliosi, *Helter Skelter: The True Story of the Manson Murders* (New York and London, 2001), p. 313.
50 Davis, *The Visionary State*, p. 184.
51 Bugliosi, *Helter Skelter*, p. 312.
52 John Gilmore and Ron Kenner, *The Garbage People: The Story of Charles Manson* (Los Angeles, CA, 1971). The quotation appears inside the title page of this original edition, but is omitted from the more recent and slightly retitled reprint: John Gilmore and Roy Kenner, *The Garbage People: The Trip to Helter-Skelter and Beyond with Charlie Manson and Family* (Los Angeles, CA, 2019).
53 Joseph Masco, 'A Notebook on Desert Modernism: From the Nevada Test Site to Liberace's Two-Hundred Pound Suit', in *Histories of the Future*, ed. Daniel Rosenberg and Susan Harding (Durham, NC, 2005), pp. 19–50 (p. 23).
54 Ibid.
55 Day and Hall, *America's Most Sustainable Cities and Regions*, p. 94.
56 Susan Lepselter, *Resonance of Unseen Things: Poetics, Power, Captivity, and UFOs in the American Uncanny* (London and Ann Arbor, MI, 2016), p. 80.
57 Valerie L. Kuletz, *The Tainted Desert: Environmental and Social Ruin in the American West* (New York and London, 1998), pp. xiv–xv.
58 Ibid.
59 Masco, 'A Notebook on Desert Modernism', pp. 28–9.
60 Kuletz, *The Tainted Desert*, p. 67.
61 Vanderbilt, *Survival City*, p. 27.
62 W. J. Hennigan, 'The New Nuclear Poker', *Time*, 12 February 2018, p. 21.
63 Mike Davis, *Dead Cities and Other Tales* (New York, 2002), p. 37.
64 Fox, *The Void, the Grid and the Sign*, p. 39.
65 Peter C. van Wyck, *Signs of Danger: Waste, Trauma, and Nuclear Threat* (Minneapolis, MN, 2004).
66 Vanderbilt, *Survival City*, p. 95.
67 Mike Davis, 'House of Cards', *Sierra Magazine*, LXXX/6 (November–December 1995), pp. 36–43.
68 John Gregory Dunne, *Vegas: A Memoir of a Dark Season* (New York, 1974), p. 21.
69 Banham, *Scenes in America Deserta*, p. 41.
70 See, for example, Jean Baudrillard, *America*, trans. Chris Turner (London, 1988); William L. Fox, *In the Desert of Desire: Las Vegas and the Culture of the Spectacle* (Reno, NV, 2005).

71 Fox, *In the Desert of Desire*, p. 3.
72 Baudrillard, *America*, p. 67.
73 Jonathan Rendall, *This Bloody Mary Is the Last Thing I Own* (London, 1997).
74 Davis, *Dead Cities*, p. 86.
75 Ibid., p. 87.
76 Rem Koolhaas, 'Junkspace', *October*, C (Spring 2002), pp. 175–90.
77 Day and Hall, *America's Most Sustainable Cities and Regions*, p. 94.
78 Ibid., p. 86.
79 Hal Rothman, *Neon Metropolis: How Las Vegas Started the Twenty-First Century* (London and New York, 2003), p. xxviii.
80 Sarah Chaplin, 'Heterotopia deserta: Las Vegas and Other Spaces', in *Intersections: Architectural Histories and Critical Theories*, ed. Iain Borden and Jane Rendell (London, 2000), pp. 203–20 (p. 215).
81 Robert Venturi, Denise Scott Brown and Steven Izenour, *Learning from Las Vegas: The Forgotten Symbols of Architectural Form* (Cambridge, MA, 1977).
82 Sally Denton and Roger Morris, *The Money and the Power: The Making of Las Vegas and Its Hold on America, 1947–2000* (London, 2002), p. 6.
83 Peter Guralnick, *Careless Love: The Unmaking of Elvis Presley* (London, 1999), pp. 447–8.
84 Baudrillard, *America*, p. 184.
85 Denton and Morris, *The Money and the Power*, pp. 3–7.
86 Martin Scorsese (dir.), *Casino* (Universal Pictures, 1995). I am quoting from the film dialogue.
87 Hal Rothman, 'Colony, Capital and Casino: Money in the Real Las Vegas', in *The Grit beneath the Glitter: Tales from the Real Las Vegas*, ed. Hal K. Rothman and Mike Davis (Berkeley, CA, and London, 2002), pp. 307–8.

Conclusion: Data Wastelands

1 Jonathan Crary, *24/7: Late Capitalism and the Ends of Sleep* (London, 2013), p. 40.
2 Bruce Sterling, 'Atemporality for the Creative Artist', *Wired* (February 2010), available at www.wired.com.
3 Ibid.
4 Michael Bracewell, *Re-Make/Re-Model: Becoming Roxy Music* (London, 2007), p. 323.
5 Harvie Ferguson, *Self-Identity and Everyday Life* (New York and London, 2009), pp. 165–6.
6 Ibid., p. 165.
7 Harvie Ferguson, 'Exteriority: Boredom, Disgust, and the Margins of Humanity', in *Aesthetic Fatigue: Modernity and the Language of Waste*,

ed. John Scanlan and John F. M. Clark (Newcastle upon Tyne, 2013), pp. 220–50 (p. 225).

8 Ibid., p. 226.

9 Fred Pearce, 'Energy Hogs: Can the World's Huge Data Centers Be Made More Efficient?', *Yale Environment 360*, https:// e360.yale.edu/features, accessed 30 November 2022.

10 Carlo Ginzburg, *Clues, Myths and the Historical Method*, trans. John and Ann C. Tedeschi (Baltimore, MD, 1989), p. 101.

11 Steven Levy, *Hackers: Heroes of the Computer Revolution*, 25th anniversary edn (Sebastopol, CA, 2010), p. 39.

12 Brand, quoted in Levy, *Hackers*, p. 384.

13 Robert Cookson, 'London's Ban for Spy Bins Highlights Lure of Big Data to Business', *Financial Times*, 13 August 2013, p. 2.

14 Ibid.

15 Thomas Pynchon, *The Crying of Lot 49* (Philadelphia, PA, and New York, 1966). The acronym stands for 'We Await Silent Tristero's Empire'.

SELECT BIBLIOGRAPHY

Ackroyd, Peter, *London: The Concise Biography* (London, 2012)
Banham, Reyner P., 'A Throw-Away Aesthetic', in *Design by Choice*, ed. Penny Sparke (New York, 1981), pp. 90–93
—, *Scenes in America Deserta* (London, 1982)
Baudrillard, Jean, *America*, trans. Chris Turner (London, 1988)
Benjamin, Walter, *The Arcades Project*, trans. Howard Eiland and Kevin McLaughlin (Cambridge, MA, 1999)
Brisley, Stuart, *Beyond Reason: Ordure* (London, 2003)
Callenbach, Ernest, *Ecotopia: A Novel about Ecology, People and Politics in 1999* (London, 1978)
Crary, Jonathan, *24/7: Late Capitalism and the Ends of Sleep* (London, 2013)
Csikszentmihalyi, Mihaly, and Eugene Rochberg-Halton, *The Meaning of Things: Domestic Symbols and the Self* (Cambridge, 1981)
Davis, Mike, *Dead Cities and Other Tales* (New York, 2002)
DeLillo, Don, *Underworld* (New York, 1997)
Dick, Philip K., *Do Androids Dream of Electric Sheep?* [1968] (Waterville, ME, 2001)
Eno, Brian, *A Year with Swollen Appendices: Brian Eno's Diary* (London, 1996)
Gehry, Frank O., *The Architecture of Frank Gehry* (New York, 1986)
Hoffman, Abbie, *Revolution for the Hell of It* [1968] (New York, 2005)
Hollingshead, John, *Underground London* (London, 1862)
McGuirk, Justin, ed., *Waste Age: What Can Design Do?* (London, 2021)
Mayhew, Henry, *London Labour and the London Poor*, 4 vols (London, 1861–4)
Melosi, Martin, *Garbage in the Cities: Refuse, Reform, and the Environment, 1880–1980* (College Station, TX, 1981)
Mumford, Lewis, *Technics and Civilization* (New York, 1934)
Nadar, Félix, *When I Was a Photographer*, trans. Eduardo Cadava and Liana Theodoratu (Cambridge, MA, 2015)
Packard, Vance, *The Waste Makers* [1960] (Harmondsworth, 1963)

Pawley, Martin, *Garbage Housing* (London, 1975)
Sabine, Ernest L., 'City Cleaning in Mediaeval London', *Speculum*, XII/1 (1937), pp. 19–43
Toffler, Alvin, *Future Shock* (New York, 1970)
Van Wyck, Peter C., *Signs of Danger: Waste, Trauma, and Nuclear Threat* (Minneapolis, MN, 2004)
Vanderbilt, Tom, *Survival City: Adventures among the Ruins of Atomic America* (New York, 2002)
White, Lynn, Jr, 'The Historical Roots of Our Ecologic Crisis', *Science*, CLV/3767 (10 March 1967), pp. 1203–7
Wilson, Elizabeth, *Adorned in Dreams: Fashion and Modernity* (Berkeley, CA, 1987)

ACKNOWLEDGEMENTS

Parts of this book have appeared in previous publications, which I would like to acknowledge. Portions of Chapter Six: 'Temporalities' were first developed in 'Scenes from a Desert Wasteland' in *Aesthetic Fatigue: Modernity and the Language of Waste*, ed. J. Scanlan and J.F.M. Clark (Newcastle upon Tyne, 2013), and appear here in substantially revised form. The sections 'Break Down' (in Chapter Two: 'Objects') and 'The Trashman' (in Chapter Five: 'Projections') are substantially expanded treatments of subjects that were touched upon more briefly in an essay titled 'Waste/Art', written for an exhibition staged by the European Parliament's House of European History in Brussels (January 2023–January 2024), which appeared in the exhibition catalogue *Throwaway: The History of a Modern Crisis*, ed. Christine Dupont, Stéphanie Gonçalves and Emma Teworte (Luxembourg, 2023), pp. 126–43. An earlier version of the section titled 'The Litter Craft' (Chapter Two) was used as part of an exhibition of jewellery art by Dauvit Alexander and Dan Russell, titled *A Waste Land*, held at Vittoria Street Gallery, Birmingham, and the Sun Pier House Gallery, Chatham, Kent (February–March 2019). I would like to thank Dauvit and Dan for the invitation to contribute to their exhibition, and James Grady for first introducing me to Dauvit many years ago. Thanks also to my former 'colleagues in waste' at the University of St Andrews: Timothy Cooper, Mark Riley and John F. M. Clark (to whom I owe additional thanks for sourcing and sharing the 1968 'Garbage' film discussed in the Introduction, many years ago. It took twenty years, but I finally found a way to fit it in).

At Reaktion Books, I would especially like to thank my editor, Michael Leaman, for his patience while I completed this long overdue book; Susannah Jayes and Alex Ciobanu, who worked with me on the images; and not least Emma Devlin, Aimee Selby, Amy Salter and other anonymous readers for going through the text with a keen eye for errors and for the many helpful suggestions that will have improved the text. It goes without saying that any remaining errors or omissions remain my own.

PHOTO ACKNOWLEDGEMENTS

The author and publishers wish to express their thanks to the sources listed below for illustrative material and/or permission to reproduce it.

Alamy: pp. 14 (PA Images), 201 (Trinity Mirror/Mirrorpix), 204 (John Heseltine); courtesy of the artist/designer Dauvit Alexander: p. 93; *Blade Runner* (1982, dir. Ridley Scott): p. 182 top and centre (Production companies: The Ladd Company/Shaw Brothers/Blade Runner Partnership); Dreamstime: pp. 24 (Fabio Pagani), 59 (Kovalenkov Petr), 143 (James Kirkikis), 146 (Truecapture), 177 (Mathiasrhode), 222 (Wirestock), 227 (Photogolfer); courtesy of Frank O. Gehry & Gehry Design, LLC: p. 140; from J. R. Green, *A Short History of the English People*, vol. III (London and New York, 1893): p. 43; © Michael Landy: p. 88 (Courtesy the artist and Thomas Dane, Gallery. Commissioned by *The Times*/Artangel Open, supported by Arts Council England. Photo: Donald Smith); Library of Congress, Prints and Photographs Division, Washington, DC: pp. 113, 116, 239, 247; Musée Carnavalet, Histoire de Paris: p. 20; courtesy of Northern Monk: p. 153; Unsplash: p. 152 (gomi); Wellcome Collection: p. 33 (CC BY 4.0); Wikimedia Commons: pp. 11 (© DACS 2025, photo Jens Cederskjold/CC BY 3.0 Unported), 35 (Gary Todd/CC0 1.0 Universal), 49 (from Henry Mayhew, *London Labour and the London Poor*, vol. II (1861)/Public Domain), 53 (first published in *To-morrow: A Peaceful Path to Real Reform* (1898), by Swan Sonnenschien & Co/Reprinted in *Garden Cities of To-Morrow* (1902)/Public Domain), 71 (Guculen/CC BY-SA 4.0 International), 73 (Self Storage by Peter McDermott/CC BY-SA 2.0 Generic), 98 (Ivan Radic/CC BY 2.0 Generic), 106 (A Disappearing Act/CC BY-SA 2.0 Generic), 111 (Ministry of Information Second World War Press Agency Print Collection/Public Domain), 121 (*Willesden Salvage* by Marathon/CC BY-SA 2.0 Generic), 125 (haymarketrebel/CC BY 2.0 Generic), 134 (Jill Clardy/Ford Galaxy – Hot Gal/CC BY-SA 2.0 Generic), 226 (BLM Nevada/Burning Man 2014: Caravansary/Photo by Dana Wilson, BLM Public Affairs/CC BY 2.0 Generic), 245 (Andrew Ferguson/CC BY-SA 4.0 International/CC BY-SA 3.0 Unported/CC BY-SA 2.5 Generic/CC BY-SA 2.0 Generic/CC BY 1.0 Generic), 256 (Gauthier DELECROIX – 郭天 from Qingdao, China/CC BY 2.0 Generic), 259 (Simon Klose/CC BY 3.0 Unported).

INDEX

Page numbers in *italics* refer to illustrations